The End of Discovery

Russell Stannard is Emeritus Professor of Physics at the Open University where for 21 years he headed the Department of Physics and Astronomy. A high energy nuclear physicist, he has carried out research at CERN in Geneva and at other laboratories in the USA and Europe. Among his awards he has the OBE, he received the Bragg Medal from the Institute of Physics, and has been made Fellow of University College London. In 1986 he was awarded the Templeton Project Trust Award for significant contributions to the field of spiritual values; in particular for contributions to the greater understanding of science and religion. His trilogy of *Uncle Albert* books introduced children of 10+ to relativity and quantum theory, and he wrote *Relativity: A Very Short Introduction* in 2008.

THE END OF
DISCOVERY

BY

RUSSELL STANNARD

OXFORD
UNIVERSITY PRESS

OXFORD
UNIVERSITY PRESS

Great Clarendon Street, Oxford OX2 6DP

Oxford University Press is a department of the University of Oxford.
It furthers the University's objective of excellence in research, scholarship,
and education by publishing worldwide in

Oxford New York

Auckland Cape Town Dar es Salaam Hong Kong Karachi
Kuala Lumpur Madrid Melbourne Mexico City Nairobi
New Delhi Shanghai Taipei Toronto

With offices in

Argentina Austria Brazil Chile Czech Republic France Greece
Guatemala Hungary Italy Japan Poland Portugal Singapore
South Korea Switzerland Thailand Turkey Ukraine Vietnam

Oxford is a registered trade mark of Oxford University Press
in the UK and in certain other countries

Published in the United States
by Oxford University Press Inc., New York

British Library Cataloguing in Publication Data
Data available

Library of Congress Cataloging in Publication Data
Library of Congress Control Number: 2010930293

Typeset by SPI Publisher Services, Pondicherry, India
Printed in Great Britain
on acid-free paper by
Clays Ltd, St Ives plc

ISBN 978–0–19–958524–3 (Hbk.)
ISBN 978–0–19–964571–8 (Pbk.)

CONTENTS

LIST OF ILLUSTRATIONS

INTRODUCTION

We take it for granted that science, by its very nature, progressively advances. But it was not always so. And more importantly, it will not continue to be so—not indefinitely. Eventually the pursuit of scientific knowledge will come to an end. Not the *applications* of science. There will doubtless always be new uses of scientific knowledge. Technology will continue. But fundamental science itself—the making of fresh discoveries as to how the world is constructed and behaves—that process, almost certainly, will at some stage grind to a halt.

I am not talking about the immediate future. As is well known, certain voices were raised at the end of the 19th century that claimed, with the formulation of Newton's laws of dynamics and gravity and Maxwell's laws of electromagnetism, that there were not likely to be any further significant advances in scientific understanding. How wrong they turned out to be! Even in our own times there have been those who have claimed that science has already come to an end. For example, in 1996, John Horgan wrote a somewhat similarly entitled book, *The End of Science*, in which he forecast that there would be no further significant discoveries. Let me be clear: I myself make no such claim. I fully expect science to make many further fundamental discoveries over the next decades, or indeed centuries. Indeed, I do not see how we shall even know for certain when we have reached the end of the road. How can one *prove* that there will be no more progress? It cannot be done. I suppose that if our descendents reach the stage where they notice that their physics textbooks have not needed updating for the past millennium, they might conclude that there were probably more productive careers for them to follow than being a research physicist! All I am saying is that we are living in a transient stage of human development—that known as the scientific age—and that at some unknown time in the future it will end.

Why? Will it be a case of having by then discovered everything—a complete understanding of the world and of ourselves? Most unlikely. No, it will be more a case of having reached the point where we will have learned all that is *open* to us to understand. Which is not the same thing. What is possible for us to know is likely to fall far, far short of understanding everything. There will remain questions that we shall never be able to answer. Three reasons come to mind why this might be so.

In the first place we do our thinking with our brains—obviously. But what are our brains? How do we come to have them? By evolution through natural selection, the brain was fashioned progressively down through the ages in response to our ancestors' need to survive. It was an essential feature of their survival kit. By using their intelligence, our ancestors found food, shelter, and a mate; they learned to avoid, or if necessary fight, predators. The brain was primarily a practical means for efficiently fulfilling those functions.

That being so, it is a bonus when the brain additionally proves successful at making scientific discoveries that have little or no relevance to those aims: knowing the universe began with a Big Bang, or that matter is made up of quarks and electrons, or that DNA is in the form of a double helix. None of these play any part in helping us to survive to the point where we can mate and pass on that DNA to our offspring. And what of the capacity to understand abstract mathematics, make music, compose poetry, and engage in a host of other activities that go to make up human culture? It is right that we should exploit these unexpected abilities to the full. But such successes should not blind us to the fact that, when it comes to seeking a deep scientific understanding of the world, we have to do our science with an instrument that evolved in response to other requirements. This surely is reason enough to be cautious about claims that it is only a question of time before it will have unlocked all of nature's secrets.

A second reason for suspecting that there might be limits to scientific endeavour has to do with practical considerations. One of the most successful branches of science in recent times has been my own research field of high energy nuclear physics. This involves studying collisions of subatomic particles accelerated to high energies in machines called particle accelerators. These are generally shaped like giant hollow circular doughnuts. The particles whirl around this racetrack many times, all the while being accelerated by electric forces. They are kept on course by magnetic fields. An analogy would be an Olympic hammer thrower whirling the ball round and round gaining momentum before release. The more energetic the particles, the more difficult it becomes to hold them on course; the strength of the magnetic field has to be increased accordingly. This is achieved by increasing the current in the electromagnets. But eventually the magnets reach their maximum strength, and that sets the maximum energy the particles can have in that particular machine. To attain still higher energies one needs more magnets. These can be accommodated only if one builds another machine—one with a bigger circumference. Currently the largest accelerator is the one at the CERN laboratory in Geneva where I used to work—the joint European centre for this kind of research. The doughnut is 27 kilometres in circumference.

Throughout the history of high energy physics, it has been found that, as a fairly general rule, each time there has been a significant step up in energy—with the introduction of a yet bigger machine—new and quite unexpected discoveries have been made. This raises the question: how big a machine would be needed in order to uncover the final piece of the jigsaw puzzle? Why should all the indispensable experimental data for formulating a final complete theory happen to match what we humans are able to achieve in practical and economic terms? A currently fashionable theory, for example, is that the ultimate constituents of matter might consist of very tiny vibrating strings. They are expected to be so tiny that, in order to check them out, it would take an accelerator the size of a galaxy! So we already know that this type of approach is not going to work.

The third reason for suspecting that the scientific enterprise will fall short of providing a complete understanding of everything is the suspicion that we can perhaps already begin to discern where some of those limits might be. These relate to questions that have been around for a long time, stubbornly defying all attempts at resolution—possibly because, for us, they are intrinsically unanswerable.

There is a phrase 'beating the bounds'. It refers to the ancient custom whereby the people belonging to a community would periodically set out to visit the various features—trees, walls, rocks, hedges—that set the boundaries marking the extent of their parish. They would pause at each landmark and strike it with a stick to confirm that this was the limit of the council's jurisdiction. In this book, we shall be engaged in a somewhat similar endeavour—trying to trace out the limits of science. I shall be introducing you to the latest achievements of science and to the problems and questions that face us today at the deepest levels of inquiry. In the process, we shall come across instances where we appear to have come up against an insuperable barrier—a limitation that might define for all time the extent of the domain of science in that particular direction.

In what follows I have, for easy reference, highlighted the major outstanding questions by placing them in the margin.

It is to be hoped that some of these will one day receive an answer. But others not. At this point in time it is not possible to be completely sure which questions are unanswerable in some absolute way. Defining where exactly the boundaries lie that separate the knowable from the unknowable is inevitably subject to uncertainty. As we go along, I shall be giving an indication as to which of these problems I personally judge to be intractable. But if subsequently I am proved wrong about some question, and it is eventually shown to have an accessible answer, I shall be delighted.

There is perhaps one final point that I should make before we get started. In drawing attention to the limitations of science, I do not wish to give the impression that I am in any way disparaging or belittling the achievements of science. Of course not. I have spent my whole working life as a scientist. I love the subject—especially my own branch which is physics. Indeed, one of the aims of this book is to share with you my own deep enthusiasm for science—an enthusiasm that began with my discovery of a book in my college library. It was dark blue, faded, and fat and it was sitting on the bottom shelf. I've forgotten who it was by. All I do know is that it was an introduction to Einstein's theory of relativity. It completely blew my mind. I had never read anything so astonishing and wonderful. It described how at high speed, time slows down. Travel fast enough and you will appear to others to live for ever. At high speed, space contracts and you get squashed up. But you don't feel a thing because the atoms in your body are squashed up and so do not need as much space to fit in. You can't travel faster than the speed of light (300,000 kilometres, or 186,000 miles, per second) no matter how powerful the thrust of the rocket of your spacecraft. This is because the faster you go, the more difficult it is to accelerate further. One interpretation of this holds that with increasing speed you become heavier. On approaching the speed of light you become infinitely heavy—and there is no way of accelerating an object that is infinitely heavy—hence the speed limit.

Heady stuff. It was for me the start of a life-long love affair with physics. Most of my subsequent career was devoted to being a high energy nuclear physicist, carrying out experiments into the ultimate structure of matter, space, and time—the really fundamental scientific questions. I look forward to introducing you, in later chapters, to these and other lines of research. In doing so, I shall assume no more than very elementary previous knowledge. When dealing with advanced topics such as relativity and quantum theory I shall be explaining, from scratch, just as much as you need to know in order to appreciate the mystery of what we have come up against.

But, you might be thinking, why engage in such an exercise? In the first place, as I have said, it gives me an opportunity to share with you some of the finest achievements of scientific thinking—the kind of science that has held me in its thrall for so long. Following through for oneself the thoughts of great scientists such as Einstein can be an exhilarating and attitude-changing experience. But in addition, it is to be seen as a call to exercise a measure of humility. We live in an age when certain scientists allow their love of the subject to get the better of them; they make extravagant claims as to the scope and power of scientific thinking. The claim is made that science is the only route to knowledge, and that ultimately it will bring us a complete understanding of everything. It is a philosophy that goes under the name of **scientism**. This book is an attempt to combat such unrealistic expectations. While seeking to promote an appreciation for the achievements of science, it is also intended to engender an even greater sense of awe when faced with the mystery of existence.

1

Brain and consciousness

We begin our tour of the boundaries with a long-standing problem. It concerns the brain: a configuration of atoms and molecules, chemical flows, and electrical currents. Some liken it to an elaborate computer. But unlike any of the computers built and designed so far, the brain is conscious—it is aware of itself. All this physical activity is accompanied by feelings and emotions—love, fear, joy, anger, pain.

Medicines and drugs alter the physical processes going on in the brain and, as a consequence, the mental experiences that go with them—from simple headache remedies and sleeping pills to hallucinogenic drugs. The passing of electric shocks through the brain has been used in an attempt to alleviate mental depression. Using various experimental techniques, neuroscientists have made great strides in identifying which patterns of brain activity are accompanied by which mental experiences. Progress in this field has been impressive.

But note, there is nothing about these physical patterns of behaviour that *in themselves* inform us that they are accompanied by someone having a mental experience. The subject has to tell us what they are thinking at the time. Even then, we have to take their word for it, there being no other way of checking it out. Once we have been told what they are experiencing it becomes reasonable to infer that, on coming across a similar pattern in another subject's brain, there will again be the same sort of mental experiences. But that's all it is: an inference.

In fact, why are there mental experiences at all? This question is the famous, or infamous, mind/brain problem—the problem of consciousness. It has been around for a very long time. Occasionally the claim is made to have solved it. The most well known of these is probably Dan Dennett's book, *Consciousness Explained.* But these attempts rarely convince anyone—other than the authors themselves perhaps.

We do not even know which things are conscious. I am conscious. I know that from direct experience. I suspect that you are too because you have a brain like mine and you talk as though you have mental experiences. Chimpanzees? Yes. Worms—maybe—to some extent. Though it has to be said when a gardener accidentally slices one in half, both halves writhe about. Are both halves in agony? Does the worm now have two minds whereas previously it had only one? Or does it not have a mind at all? How about bacteria? The Sun? Certainly not the Sun. But where was the dividing line? How could we ever find out?

Suppose we were to build a computer—a supercomputer of a complexity comparable to that of the human brain. Not a realistic prospect perhaps, but just suppose. And this computer were to declare 'I am conscious', would we believe it? It sounds a reasonable assumption, but such a declaration would not constitute proof. There would remain a nagging worry that, just as we teach a child how to speak, we had to teach the computer the rudiments of language. In that process might we not have programmed the computer to make such a declaration? Just because my sat nav talks to me like a knowledgeable travelling companion does not mean that it is experiencing the journey in the way that I am able to. It might well sound put out when I decide to take a different route from the one it has suggested, but that is not to say that I have really hurt its feelings—or that it has any feelings at all.

The problem of consciousness has been around for centuries, indeed millennia, and during that time has taxed the finest philosophical minds. In calling it a 'problem', we are presumably referring to the fact that we have two ways of talking: a scientific language where we use terms like *brain, atoms, chemical flows, electrical currents,* etc., and a psychological way of talking based on concepts such as *mind, emotions, love, fear, pain, making a decision,* etc. Although there can be correlations between these two ways of talking (a particular pattern of brain activity being associated with the psychological response of feeling pain, for instance), we are nevertheless stuck with using two entirely different types of language. There is no way we can see concepts such as *love, fear,* or *pain* being quantified and figuring in a physics equation alongside terms such as *mass* and *electric charge*. Having two languages strikes us as messy. One instinctively feels that a 'proper' explanation of what is going on should be able to combine all the information into a single, coherent, universal language. But this seems unattainable.

I reckon the answer is that we have to revise our understanding of what constitutes a 'proper' explanation. Certainly we know that each language contains information and insight that is not to be found in the other. Each language on its own is impoverished and cannot provide a comprehensive understanding as to what it is to be a human being. Perhaps a proper, full, understanding consists of holding in the mind these two separate languages, emphasising neither at the expense of the other.

At any rate, it seems to me that this problem is our first encounter with a question that is unlikely ever to be solved to everyone's satisfaction.

The problem of ≺
consciousness.

The second is closely related to the first. We have seen how patterns of brain activity are associated with mental experiences. It seems reasonable to assume that for any given brain activity, there will be an exactly corresponding mental experience, such that if the physiological pattern is repeated, the same mental experience will be repeated. This is not known for certain but appears to be a good working hypothesis. As regards the workings of the brain, these will be governed by the laws of physics. This means that for a given physical state of the brain at a given point in time, and from a knowledge of all external influences upon it, one ought to be able to predict what the succeeding physical state will be from a simple application of those laws. We say that the future state is determined.

But surely this implies that if we know all about which mental states correspond to which physical states of the brain, it should be possible to predict from one mental state what the next one will be. The future mental state will be that corresponding to the future physical state—which, as we have just seen, is determined. What does this do to our sense of free will—the idea that the future is open and will be dependent on whatever choice we make between a set of possible alternative courses of action? Surely, so goes the argument, if some neuroscientist were able to identify exactly what the state of my brain is at some particular point in time—for example, while I am 'making up my mind which of two pencils I shall pick up from the desk'— he or she could predict what the future state of my brain will be after I am supposed to have made a 'decision', and hence could identify what my mental choice will be? In other words, there is no real choice. The notion that the future is open is but an illusion.

This is the free will/determinism problem. It is another problem scientists and philosophers have wrestled with for a long time—the problem of how to reconcile the grinding predictability of the physical brain with the mental sense that the future is open and it is up to us to decide what it will be.

Some scientists have sought to solve the problem by calling upon quantum theory. Quantum theory is necessary for the description of the behaviour of very small objects—atoms, molecules, subatomic particles. We shall say much more about this later on. But for the time being all one needs to know is that on the very small scale—the level of molecules, atoms, and subatomic particles—the future is *not* predictable. There is an inbuilt uncertainty. Given a particular situation it is impossible to be sure what will happen next. All one can do is predict the odds of various possible outcomes. This is summed up in something called **Heisenberg's uncertainty principle**—named after the German physicist who first discovered this fundamental principle. Thus, we find that the future is *not* strictly speaking determined. Rather it is governed by chance—random chance.

This raises an intriguing possibility: if the action of the brain corresponding to the making of a choice—which pencil to pick up, for instance—if that action is something happening on the very small scale characteristic of the movement of individual atoms, then the behaviour is governed by chance; it is *not* predetermined. Some have latched on to this to claim that this is how we come to make a free will decision.

But is this the answer? Suppose I cannot make up my mind which pencil to pick up. What can I do? Toss a coin.

I leave it to chance. But in doing so, am I making a decision—a conscious decision? No. Indeed, I am *opting out* of making a decision. The use of chance is an alternative to making a conscious decision.

A further idea is to adopt a dualistic approach, according to which the mind and brain are considered to be different entities such that not only can the brain affect mental states, but the mind can influence the brain processes. In other words, the matter that makes up the brain does not slavishly follow the laws of nature; the mind (whatever kind of entity that is supposed to be) can somehow intervene and override nature's normal course. A variant of this argument is to assume that the mind does nothing that obviously alters the course of events according to the laws, but instead makes its mark by subtly altering the odds on the possible outcomes allowed by quantum uncertainties—this being done in such a way as not to violate the uncertainty principle.

Yet another approach is to argue that free will and determinism are compatible with each other—one does not have to choose between competing alternatives. This suggestion holds that too often the idea of determinism is coupled to the notion of compulsion against one's will—that if we were subject to determinism, our mental life would be one in which we would feel ourselves constantly fighting a losing battle against inexorable forces compelling us so to act, regardless of our own intention. As a result, we would find ourselves having to conform to some preordained plan that was not of our own devising. But this does not necessarily follow. According to this viewpoint, determinism is perfectly compatible with our invariably acting according to our will. In all situations, what our physical body is about to do is perfectly in tune with what our mind wills to do in that set of circumstances. It is not so much that we are *free* to act according to our will, as *guaranteed* so to act. The physical necessity coincides with what we mentally experience as the authentic expression of our true nature. We are being ourselves. Why would we want a 'freedom' to act otherwise?

A particularly notable contribution to the free will/determinism debate was made in 1983 by Benjamin Libet and his collaborators. They performed an experiment whereby a subject had to decide when to flex their wrist, and then report the exact time at which that decision was taken. A scan revealed that brain activity associated with this 'decision' occurred 300 to 500 milliseconds before that time. This result appeared to show that the course of action was actually determined significantly earlier than the conscious experience of 'making a decision'. The experiment was repeated by Chin Siong Soon in 2008 in a version where the subject had to decide whether to press a button with their left or right index finger. It was found that the nature of the decision could be found encoded in brain activity up to ten seconds before the subject reported that they had made their choice.

Results like these might appear to some to prove that there is no such thing as free will. All the conscious mind is doing is giving expression to what has earlier been fixed by the physical processes going on in the brain. The conscious mind is not playing any part in deciding what the course of action should be.

But such an interpretation might be too hasty. After all what is it that constitutes the whole thinking person who is supposed to be responsible for making the decision—a decision that reflects the nature and character of that person? It is what psychologists call the **psyche**. It is the combination of the conscious and unconscious self, the latter being the complex repository of memories, genetic influences on behaviour patterns, and so on. Conscious thinking and actions might be thought of as but the outward manifestations of that whole person. The brain processes that precede the conscious pressing of the button, or whatever, might well be regarded as merely the physical correlate of the decision being made by the unconscious component of the psyche.

None of the ideas about free will have gained universal acceptance, and it is difficult to see how the dilemma between free will and determinism is to be resolved to everyone's satisfaction. Some will conclude that we are not as free as we might wish to think we are, while others will continue to side with Dr Johnson, who declared: 'Our will is free, and there's an end on't'.

➢ *The free will/ determinism problem.*

17

2

Creation of the cosmos

T he task of science is to explain the world—what it consists of, how it operates, and how it came to be the way it is. This will involve us in examining both its most minute detail and also its broad overall architecture. We begin with the latter—the cosmos.

We live on planet Earth, orbiting the Sun. The Sun is a star much like all the other stars. Stars are gathered together into great swirling whirlpools called **galaxies**. Our galaxy is called the Milky Way Galaxy. Galaxies in their turn are gathered into **clusters of galaxies**.

On observing distant clusters we find that they are receding into the distance away from us. The further off a galaxy cluster, the faster it is receding. A cluster that is twice as far from us as another is receding at twice the speed; ten times the distance, ten times the speed. This is called the **Hubble expansion**, after Edwin Hubble, who discovered the expansion of the universe in 1929. From the speeds of the clusters and their distances from us we are able to calculate that at a distant time in the past (13.7 billion years ago to be precise) all the matter of the universe was together at a point. There was a great explosion—the **Big Bang**—and this flying apart of the clusters is the aftermath of that event. This conclusion is backed up by independent evidence. For example, such a violent explosion should have been accompanied by a hot fireball. Sure enough, the cooled down remnants of that fireball were detected in 1965 by Arno Penzias and Robert Wilson; it is called the **cosmic microwave background radiation**. Not only that, but it is possible to calculate what chemical composition one would expect for material emerging from the Big Bang. It should consist mostly of the two lightest elements, hydrogen and helium in the ratio three to one by mass, and very little besides. That indeed proves to be the observed composition of the **interstellar gas**—the raw material out of which the stars later formed. In the face of these corroborating indications, few people today doubt that the Big Bang theory is correct.

It was such a cataclysmic event that it appears natural to assume that it marked the instant at which the universe came into being. In which case, it further seems natural to ask what caused the Big Bang?

In trying to answer that question we need to probe ever closer to the instant of the explosion, and indeed beyond that instant to what might have preceded it. How might we do that—given that it all happened a long time ago? One method, which appears from the outset promising, is to look deep into space. When we observe a distant galaxy cluster, we are not seeing it as that object is today, but how it was when it emitted the light that we are at present receiving. Light travels at a finite speed so it takes time to reach us. Thus, looking at distant objects is equivalent to looking back in time; the further away the object, the further back in time we probe. So how far back in time can we go? Unfortunately, no further than 300,000 years after the Big Bang. Beyond that we encounter a radiation fog. The trouble is that the Big Bang was so violent that all that came out of it was radiation and subatomic particles. It was not until things cooled down so as to allow the subatomic particles to come together to form atoms that the radiation could be absorbed and the universe became transparent to light. And that did not happen until 300,000 years after the Big Bang.

However, there is a possibility—admittedly a *very* slim possibility—that one might be able to do better. Not only do we encounter light and other forms of electromagnetic radiation in the universe but also **neutrinos**. Neutrinos are a kind of fundamental particle well known for hardly ever interacting with anything. One could pass a typical neutrino though the Earth from one side to the other 100,000 million times before it had a 50:50 chance of hitting anything. But experimental techniques have been developed which have detected them. This holds out the prospect that we might be able to detect neutrinos that penetrated the initial radiation fog and could therefore reveal what was happening at earlier epochs. This way one could, in principle at least, probe back to about one second after the Big Bang. But the practical difficulties involved in achieving this are enormous. Beyond that, there is the even more remote possibility of using **gravitational waves**—a type of wave generated when massive bodies are accelerated. Theoretically, gravitational waves could probe right back to the instant of the Big Bang. Unfortunately, they are yet more difficult to detect. Despite attempts having been made for many years, to date no-one has succeeded in detecting a gravitational wave.

➢ *How close to the instant of the Big Bang are we likely to be able to probe?*

Thus practical difficulties set limits on how close we can get to the instant of the Big Bang by direct observation. But that is not the end of the story. There are other clues—features of today's universe that owe their origins to what might have happened at earlier times. Take for example the homogeneity and isotropy of the universe. By this we mean that the universe looks much the same in whichever direction we look. The microwave background radiation, for example, is essentially uniform across the sky. Regardless of direction, it has the same spectrum of wavelengths, indicating the same temperature. This is odd. Normally one would expect things to have the same temperature only if they have been in contact with each other for a while and have had a chance to equalise their temperatures. But if everything was flung out of the Big Bang instantaneously, as we have so far been assuming, there would not have been any chance for this equalisation to take place.

In 1980, Alan Guth proposed the idea that initially there was a comparatively quiet time immediately following the instant of the Big Bang, during which there were causal connections between the universe's contents, and the equalisation of temperature was able to take place. After this, expansion set in. However, if this expansion was simply the Hubble expansion that we see going on now, the galaxy clusters would not be where they are today. What is required is that the relatively quiet period was followed by a brief period of exceedingly rapid expansion—a process we have called **inflation**. When I say 'brief', I mean *brief*. It is estimated to have begun 10^{-36} of a second after the instant of the Big Bang. That's a 1 with 36 zeros on the denominator: $\frac{1}{1,000,000,000,000,000,000,000,000,000,000,000,000}$th of a second Inflation lasted for 10^{-32} of a second before settling down to the more stately Hubble expansion that we see today.

So, inflation neatly solves the problem of the isotropy of the microwave background radiation. In addition, as we shall be explaining later, it also offers an explanation for the observed energy density of the universe. Furthermore, it should be pointed out that, although the background radiation is to most intents and purposes uniform, there are some very small-scale inhomogeneities with amplitudes of typically 10^{-5}. These must have arisen because of early inhomogeneities in the density of the primordial material—small fluctuations in density which, over the course of time, were enhanced by gravity and eventually led to the formation of galaxies. The study of these irregularities over different angular scales should throw light on how galactic structures formed.

I think you will agree that to be able to speak with reasonable confidence about was happening so soon after the Big Bang is a remarkable achievement. Although it is not possible to gain direct evidence of inflation, the indirect inferences are powerful enough to convince most scientists working in the field that inflation did indeed take place. Mind you, it should be pointed out that since Guth put forward the first theory of inflation, others have been proposed. A study of the angular distribution of inhomogeneities in the background radiation might be able to distinguish between the rival inflation theories, but this is not certain.

> *Can we be sure that inflation took place?*

> *If so, how are we to choose which type of inflation it was?*

In any event, it is important to emphasise that there is all the difference in the world between saying what the universe was like a tiny fraction of a second *after* the instant of the Big Bang, and saying what it was like at the *instant* of the Big Bang. This is because, at that instant, one presumes that all the matter of the universe would be at a point—a place of no volume—so the density would be infinite. We call this a **singularity**—and there is no way that our physics can handle such situations. It is because of this that we have no hope of extending our investigation *through* that instant to what might have preceded it—to what might have *caused* the Big Bang. Indeed, according to Stephen Hawking, we might be wrong in assuming there was a singularity. He speculates that time itself might change its nature as one imagines approaching the supposed 'first instant'. In effect, time 'melts away' and there is no first instant. That would be another possible reason why the question 'What caused the Big Bang?' will never be answered.

Was there a ◄
singularity at the
instant of the Big
Bang?

But there might be an even stranger reason. This comes about when we enquire what *kind* of explosion the Big Bang was. When we normally talk about an explosion, we have in mind something that happened at a particular position in space—a stick of dynamite placed in a hole drilled into the base of a chimney stack due for demolition, for instance. The explosion starts from a localised region and spreads out from there to fill up the rest of the surrounding space. But with the Big Bang, it was different. There was no surrounding space. Not only were all the contents of the universe squashed up, but all of space was squashed up too. It is, in fact, the expansion of space that carries the galaxy clusters apart.

24

As an analogy, consider small coins glued onto the surface of a rubber balloon. Now blow into the balloon. The coins separate. This is not because they are sliding over the rubber into regions where previously there had been no coins. It is due to the rubber itself expanding and carrying the coins along with it. That is how it is with the cosmos. The galaxy clusters are being swept apart on a tide of expanding space.

One should perhaps explain that just because space expands, that does not mean that everything is expanding in size: atoms, the Solar System, the galaxies themselves. Were that to be the case we would be unaware of such an expansion. (Measuring an expanded distance with a ruler that has expanded by the same ratio, we would get the same reading.) Though the expanding space is trying to make the constituents of atoms, the Solar System, and the galaxies drift apart, the physical forces, electrical and gravitational, which hold them together prevail. It is only when we consider the distances separating galaxy clusters that the gravitational attraction is sufficiently weak for the expansion of space to take over.

Now, if this is the first time you have come across this sort of thing, I can imagine your confusion. Space—*empty* space—must surely be just another name for nothing. That being so, how can nothing push a heavy thing like a galaxy cluster?! The truth of the matter is that physicists do not regard space as simply nothing. It is an interesting *something* with lots going on in it. I shall have more to say about this later. But for the time being it is sufficient for our purposes simply to point out that space probably started out as a point—something with no volume, and this is why we say space came into existence at the instant of the Big Bang.

And not just space, but time also. This is because (as again we shall see later) there is a very, very close connection between space and time—to the extent that one cannot have space without time, nor time without space. So if the instant of the Big Bang marked the coming into existence of space, it also marked the coming into existence of time. Thus we conclude there was no time before the Big Bang.

Now, for those seeking a *cause* of the Big Bang this raises a problem—an insuperable problem. Cause is followed by effect. Boy throws stone—cause; followed by window breaks—effect. The effect in the case we are considering is the Big Bang. So, the cause of it must have existed beforehand. But where the Big Bang is concerned, there is no beforehand. Accordingly, it is not that the singularity we came across prevents us extrapolating through the instant of the Big Bang to what happened earlier. And for that reason, we shall never know the answer to the question, 'What caused the Big Bang?' The problem appears to go deeper—much deeper. The *question itself* almost certainly has no meaning. It *sounds* a perfectly fair question. But it is not.

Does it make sense to enquire into the cause of the Big Bang? ≺

So much for the search into the origins of the universe—the mechanics of how the universe came

26

into being. But there is a yet more profound question—one that has taxed the minds of philosophers and theologians for the past three or four thousand years: Why is there a world at all?

If nothing existed, or had ever existed, would that call for an 'explanation'? No. Why should anything exist? But as soon as something exists—a universe—then the questions begin. What is responsible for it being in existence? Why is it this kind of universe rather than some other? After all, we can all dream up alternative universes; science fiction writers do it all the time. But these other universes don't exist; they remain imaginary. So, why is this one an exception? Once in existence, does it take some kind of agency to keep it continuously in existence? Does it make sense to think in terms of a 'ground of all being', as the theologian Paul Tillich put it? Or are such questions meaningless? One thing seems fairly certain, science has nothing to say on the subject. Science takes the existence of the world as a given. It is solely concerned with exploring the nature of the world it confronts. Any enquiries as to why there is something to examine in the first place is simply beyond its remit. Whether this is the point at which other types of thinking—philosophical or religious—take over becomes an open question.

➤ *Why is there something rather than nothing?*

3

The laws of nature

The world is run according to laws—the laws of nature. But why? Why are there any laws of nature? If the world had been chaotic—if it was a free-for-all—anything and everything happening, then that would not have called for explanation. But the world is not like that. As Einstein once put it, 'The most incomprehensible thing about the world is that it is comprehensible.' Where did the laws come from? Why are they *these* laws rather than some others? Why are some behaviours allowed and others not?

One sometimes hears scientists claiming that once we have a complete knowledge of the laws we shall discover that these laws are in fact the only set of laws there could possibly be—hence no mystery. But this cannot be. There can be no such final justification. What emboldens me to make such an assertion? Simply this:

Galileo once wrote, 'The book of nature is written by God in the language of mathematics.' The laws are expressed mathematically. A mathematical structure, or system, consists firstly of a set of axioms, or assumptions. These might, for instance be the axioms of Euclidean geometry—the kind of geometry we learned at school. These include the definition of a straight line as the shortest distance between two points, parallel lines not intersecting, non-parallel lines intersecting at only one point, and so on. Then built upon these axioms are the theorems. These are the consequences, or true statements, that can be deduced as a result of having accepted this particular set of axioms. I suppose the best known of these is Pythagoras's theorem regarding right-angled triangles: 'The square of the hypotenuse is equal to the sum of the squares of the other two sides.'

In trying to formulate the ultimate **Theory of Everything**, as it is sometimes called, what one is essentially doing is trying to identify the appropriate mathematical system. The axioms of that system will correspond to the laws of physics, and the theorems will be all the possible behaviours that can take place in the world as a result of the operation of, and interaction between, those laws—in other words what we see going on in the world.

At the present time we do not know which mathematical system is the right one. But for our purposes that does not matter. *All* mathematical systems have something in common, namely that from within a mathematical system there is no way of justifying the original choice of axioms; the axioms are simply a given. (Of course, if that system is but a small component of a larger mathematical system, then it might indeed be possible to justify those axioms; they might arise as provable theorems of the larger system. But such a justification comes from *outside* the smaller system of interest. And in any case the larger system will be based on its own axioms—and these are not justified.) We ourselves, being part of the world, will be modelled by a part of the mathematical system corresponding to the Theory of Everything. So we are within the mathematical system. It seems reasonable therefore to conclude that we have no possibility of justifying the underlying choice of its axioms—meaning that we have no possibility of justifying the choice of the laws of nature.

Where do the laws of nature come from? ◄

Indeed, the situation is even worse than that. There is a famous theorem called the **incompleteness theorem**. It was formulated in 1931 by Kurt Gödel. It demonstrates that within a mathematical system such as those we have been discussing, there will be statements that are true but can never be *proved* to be true. In other words, mathematics is not as rigorously complete as was previously assumed. And that in turn implies that there is bound to be a fundamental incompleteness in our understanding of the world.

Can one prove mathematically that science will be for ever incomplete? ◄

All this talk of the role of mathematics, in itself raises a teasing question: What is the status of mathematics—mathematics itself? Is it something

30

we invent—like any other language—English, French, German? Or does the logic of it all have some form of independent existence—something we discover, rather than invent. Suppose, for instance, that there were no triangles—no physical triangles because there was no physical world. Would there still be such a thing as Pythagoras's theorem? This is an issue that has been kicked around for ages—without producing any agreed consensus.

> *What is the status of mathematics?*

Finally we should mention that there has been speculation as to whether the laws of nature remain the same over time. The laws incorporate constants that govern such things as the strength of the gravitational force, the strengths of electric and magnetic forces, etc. But are these so-called 'constants' really constant—unchanging over time? This was a question originally raised by Paul Dirac in 1937. From time to time one hears reports of evidence for there having been changes to, for example, the speed of light and the **fine structure constant**. The latter governs how light interacts with charged particles, such as atomic electrons. One investigation involved gas clouds so distant from us that the processes we observe today going on in that cloud were actually taking place 12 billion years ago. The investigators examined how those gas clouds were absorbing light 12 billion years ago, and they thought that they could tell that things were different then to what they are now. This suggested that the value of the fine structure constant had changed over that period. This result was received with some scepticism, and it remains—at least for the time being—an open question as to whether any of the physical constants change with time.

Do the physical constants change with time? ≺

4

The anthropic principle

I t is a very strange world that we inhabit. It had to be
strange for us, and for other forms of life, to be able to
live in it.

For a start, consider the violence of the Big Bang. It was
exceedingly violent. But it had to be that way if we humans,
and possibly other forms of intelligent life, were eventually to
put in an appearance. Had the violence been somewhat less—
only a little less—then the mutual gravity operating between
the galaxy clusters would have got such a secure grip that the
clusters would have slowed down to a halt. With gravity still
operating, they would then have been inexorably drawn
together, all ultimately ending up squashed to a point again.
This is the so-called **Big Crunch** scenario. Moreover, all this
would have happened in a shorter time than 13.7 billion
years—the time needed for evolutionary processes to produce us. So, a less violent Big Bang would result in no life.

Nor can one afford to have the violence of the Big Bang any more violent than it was. Were that to have been the case, the gases from the Big Bang would have been expelled so fast that they would have been dispersed into the depths of space without having had sufficient time to clump together to form embryo stars. Without stars, i.e. suns, there can be no life. So, what we find is that, in order for there eventually to be life, the Big Bang violence had to be just right. The window of opportunity was exceedingly narrow. And yet our universe managed to achieve this.

How are we to explain this? A partial explanation could be provided by the inflationary scenario—that brief but important period of exceptionally fast expansion that the universe underwent immediately after the instant of the Big Bang. It is a feature of this theory that it sets up the conditions for the onset of the normal Hubble expansion. And these conditions in turn lead to a universe that is ultimately expected to slow down to a halt, but only in the infinite future. This therefore safeguards the universe from the Big Crunch possibility. So, that at least solves half the problem: the nature of the Big Bang was such that it could not have been too weak for life. But at what price? We have to invoke inflation as the 'safety measure'. But who ordered *that*?

The next point to consider is the strength of the gravitational force. This is governed by the value of a physical constant denoted by the letter G. The force of gravity is responsible for gathering together the gases emitted from the Big Bang to form high density concentrations. The gases squash down, and in the process get hot—in much the same way as air in a bicycle pump gets hot as it is compressed. If sufficient gas has been initially gathered, then the temperature rise can exceed about a million degrees. At this point the movements of the subatomic particles become so pronounced that the nuclei of hydrogen and helium (which came out of the Big Bang) can crash into each other so violently that they fuse together to form the nuclei of heavier atoms, with the release of large quantities of nuclear fusion energy—the kind of energy generated in hydrogen bombs. The Sun, and the other stars, are in effect, hydrogen bombs going off in a steady continuous manner, rather than all at once.

This is a condition very hard to achieve in a laboratory. Determined efforts to harness nuclear fusion power for peaceful purposes have so far yielded no contributions to our energy needs. But the Sun does this effortlessly. It has already burned at the present steady rate for five billion years and will continue to do so for the next five billion years. It has established a delicate balance between, on the one hand, the rate at which gravity feeds the fuel into the 'furnace' situated at the very heart of the Sun, and on the other, the rate at which the nuclear reactions proceed. Had G been somewhat greater, a larger amount of gas would have been drawn together to form a more massive star. Such a star would have reached even higher temperatures causing the burning processes to accelerate, reducing the active life for the star to only a million years. But in order for life to evolve on planet Earth, we needed a medium-sized star like the Sun, which could yield steady warmth over a period of 4.5 billion years—the time that has elapsed since the formation of the Earth. What we are saying therefore, is that if the value of G had been greater, only the more massive type of stars would have formed—the short-lived ones known as **blue giants**— and there would have been no time for life to evolve on nearby planets.

How about making the value of G less? That does not work either. The trouble with that is that when the gas collects together in clumps after the Big Bang, a reduced gravity force operating between the gas particles would not have been sufficiently strong to collect enough gas to produce a temperature rise sufficient to light the nuclear fires. No nuclear fires would mean no stars, and no stars would mean no life.

So, in order for there to be life, the force of gravity must be neither too strong nor too weak. The range of permissible values for its strength is narrow. And yet, that is where the gravity of the actual universe lies. Are we to regard this, like the correct thrust of the Big Bang, as just another coincidence?

But we haven't finished. Next we must turn our attention to the materials from which the bodies of living creatures are built. Out of the Big Bang all we got were the two lightest gases—hydrogen and helium—and precious little besides. And it *has* to be that way. Remember that we need a violent Big Bang to stop the universe from collapsing back in on itself prematurely. And because of that violence, only the lightest nuclei could survive the collisions occurring at that time; anything bigger would have been smashed up again soon after its formation. But one cannot make interesting objects like human bodies out of just hydrogen and helium. So the extra nuclei—those that go to make up the 92 different elements found on Earth—must somehow be manufactured *after* the Big Bang. That is where the stars have another important role to play. Not only do they provide a steady source of warmth to energise the processes of evolution on neighbouring planets, they have also to create the heavier atomic nuclei by fusing the lighter elements.

This is not without its difficulties. Perhaps the most important atom in the making of life is that of carbon. In a sense it is an especially 'sticky' kind of atom with plenty of 'hooks' with which to attach to other atoms. It is thus very good at cementing together the large molecules of biological interest. But forming a nucleus of carbon is by no means easy. Essentially, it consists of fusing three helium nuclei together. This is as unlikely as having three moving snooker balls colliding simultaneously. Without my going into the details of how this comes about, let me just say this: Basically, how big one subatomic particle looks to another depends upon their approach speed. At certain special approach speeds, the particles can look exceptionally large; the chances of them now colliding are correspondingly higher. This is called a **nuclear resonance**. What happens in stars is that two colliding

37

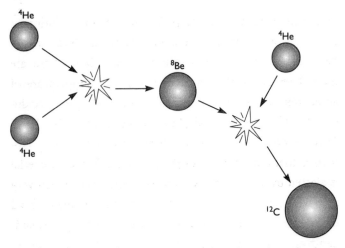

FIG 1 The fusion of three helium nuclei to form the nucleus of carbon.

helium nuclei stick together for a fleeting moment to form an unstable version of beryllium (see Figure 1). Normally, one would expect this to break apart before a third helium nucleus has time to collide with it to form a carbon nucleus. Except that, at the approach speeds to be found in stellar interiors, there just happens to be a nuclear resonance whereby the beryllium nucleus looks enormous to the approaching third helium nucleus; fusion is thus able to take place with the production of carbon. It is another of the fortunate circumstances that played a key role in making our eventual appearance on the scene possible.

We have our precious carbon. A collision between some of these carbon nuclei and further helium nuclei yields oxygen—another vital ingredient for life—and so on.

Does this mean that the stage is now set for evolution to take over and convert these raw materials into human beings? Not so. We have our materials, but where are they? They are in the centre of a star at a temperature of about ten million degrees. Hardly an environment conducive to life. The materials have to be ejected. But how? There is no rocket to propel them into space. What happens is that a proportion of the newly synthesised material is ejected by supernova explosions. These occur when very massive stars run out of fuel. They can no longer hold themselves up against their own internal gravity and suddenly collapse in on themselves. But that raises a problem. How can an *im*plosion produce an *ex*plosion? This was a conundrum that exercised the minds of astrophysicists for many years. In the event, the mechanism turned out to be the strangest imaginable. The material is blasted out by neutrinos. As we have already mentioned, neutrinos are incredibly slippery. And yet it was neutrinos that blasted out all the material that was eventually to be incorporated into our bodies. How fortunate they were not any *more* slippery than they actually are!

The material is now out among the interstellar gases. In time, this collects together to form a dense cloud, which squashes down to form another star. Outside the star there can be secondary eddies that settle down to form planets. It is now possible to have rocky planets like Earth, Mercury, Venus, and Mars. For the first generation of stars this had not been the case because, at that stage, there had only been hydrogen and helium around. Given a planet at a reasonable position away from the star for a temperate climate to prevail, one has now at last got a chance of life evolving from the inanimate chemicals lying around on the surface of the planet—what has come to be called the 'primordial soup'. How likely this is to happen is not known. As a physicist you tend to be impressed by the vast number of planets there must be out there—in other words you are impressed by how many attempts we are allowed to produce intelligent life. On this assessment, we are indeed home and dry. If on the other hand you are a biologist, you might be more impressed by the size of the hurdles that have still to be negotiated on the way to intelligent life—like for example the formation of the first complex cell, or the formation of the first multicellular organism. You might therefore be inclined to think that there must be some more 'coincidences' to follow—biological ones this time rather than the physical ones that we have been considering.

Take for instance the copying process whereby the DNA codes, containing hereditary information, are passed on from parent to offspring. Obviously this has to be done pretty accurately; if it were not, the codes would not be preserved from one generation to the next. But the copying must not be *too* accurate. Evolution by natural selection requires that there be mistakes in the copying process such as will give rise to new codes leading to novel characteristics that can then be worked upon by natural selection to sort out the advantageous variations from those that are not. If the DNA copying were done as accurately as my computer making a back-up copy of the text of this book, there would be no variations upon which natural selection could work. Thus, the accuracy of the copying process has to be poised between being too error-prone on the one hand, and too exact on the other. Is it yet another coincidence that the actual process manages to strike a happy balance?

The sum total of all the coincidences that have led to the actual universe being hospitable to life goes under the name the **anthropic principle**. It is impossible to put a hard figure on the likelihood of getting a universe suitable for the development of life from simply throwing together a bunch of physical laws at random—laws incorporating arbitrary values for the various physical constants. In talking for example about the strength of gravity having to lie within a narrow range, it is impossible to be more quantitative than that unless there is some way of specifying a permissible range of values that the strength could conceivably have taken on. If it could have been *any value whatsoever*, then the finite range would be divided by infinity, and the chances would be virtually zero.

All of which adds up to a very strange universe. How are we to account for it being so life-friendly? And it is no answer to say, 'Of course it has to be life-friendly; if it were not, we would not be here asking the question!' Such a statement is manifestly true, of course, but it goes no way towards explaining why we are here in the first place to ask the question.

41

Why is the universe ≺
life-friendly?

One possible answer is to assume that it was deliberately designed that way. The cosmologist and one-time atheist, Fred Hoyle, on making his discovery of the fortuitous nuclear resonance that leads to carbon production in stellar interiors, was moved to declare 'a commonsense interpretation of the facts suggests that a super intellect has monkeyed with the physics.' But this, of course, takes us out of science and into the realm of theology, and thus beyond the remit of this book.

Instead we note that a universe specifically designed as an environment suitable for the development of life is not the only possibility. Some have tried solving the problem by suggesting that our universe might not be alone. There might be other universes distinct from ours. Lots of them. Perhaps an infinite number of them. The whole ensemble has been called the **multiverse**. And the idea is that the laws of nature are different in all the individual universes. That way no choice has to be made. Although the vast majority of universes will be hostile to life because one or more of the conditions for the development of life were not met, sooner or later a universe will turn up that just happens—purely by chance—to have all the right laws and the right set of conditions. We being a form of life ourselves must, of course, find ourselves in one of these freak universes. The multiverse idea is clearly a possibility. But how would we ever prove or disprove that there are these other universes? Other universes, by definition cannot be contacted from our universe—otherwise they would be part of our universe.

A variant of this is that there is essentially just the one universe, but it is divided up into different domains, the laws of physics being different in each domain. According to certain versions of inflation theory, this breaking up into domains is something that could have happened during that brief period of hyperinflation occurring a fraction of a second after the instant of the Big Bang. Such an alternative would appear to hold out the prospect that if we were to travel far enough out into space we might come across the boundary separating our domain from its neighbouring domain. On crossing that border, we would find ourselves in a different kind of world run on different lines from our own. Unfortunately, the same theory as puts forward this possibility also would have us believe that these domains are so colossal in size that there is virtually no practical possibility of us ever travelling far enough to reach the border. Our entire observable universe would take up only the tiniest fraction of one of the domains.

> *Are there universes other than our own?*

Thus it would appear that the anthropic principle poses a problem that is likely to remain with us.

5

The size of the cosmos

When thinking of the cosmos one cannot help but be impressed by its sheer size. The clusters of galaxies stretch out into space as far as one can see. So, our next question is: how big is the universe?

Here we have to make a distinction between the universe and the *observable* universe. We see something by the light it sends to us. As already noted, it takes time for the light to get here. The distances involved are so immense that even travelling at 300,000 kilometres per second, it takes four years for light to reach us from even the nearest star. It takes 100,000 years to cross from one side of the Milky Way Galaxy to the other. The fact that the universe came into existence 13.7 billion years ago means we can see only those objects such that the light from them could reach us in less than 13.7 billion years. This defines the extent of the observable universe. Beyond the observable universe lies the rest of the universe—presumably. How far does *that* go on? For ever. The universe is infinitely big. That is the current thinking.

But what justification can there be for saying that? Is it because we are incapable of imagining what it would be like to come to the edge of the universe? If we did come to an edge, what would lie beyond the edge? Nothing. But wouldn't that look like empty space and, as such, wouldn't it be part of the universe—just with nothing in it? So we wouldn't have come to the end of the universe. That's why we say it goes on for ever—the universe is infinite.

An alternative solution to the problem is the idea that the universe might be closed. It would be finite in size, but have no edge. How could that be? Imagine a fly crawling over a rubber surface. It carries on in the same direction. All the time it is thinking that it is getting further and further away from its starting point. It assumes that the surface just goes on and on for ever. But then, to its surprise, it finds that, without ever changing direction and retracing its steps, it is back where it started! How is that possible? We who have been observing the fly know the answer. From our bird's eye view we are able to see that the rubber sheet is not flat; it is in fact a large balloon. The fly has simply gone right round the balloon to where it started.

In the same way—so this theory goes—it might be that if we were in a spacecraft and took off from the north pole and continued going vertically upwards—always keeping to the same direction—we might find that eventually we approached Earth from the opposite direction and landed on its south pole. That way the cosmos would have a finite size (our complete journey across the cosmos took a finite time) but there was no edge. Such a possibility would require that our space—our three-dimensional space—was somehow curved back on itself, this curvature perhaps being in some unobservable additional dimension in the same way as the fly's two-dimensional rubber surface was curved back on itself in an unobserved third dimension.

If you are trying to form a mental image of such a curvature of three-dimensional space, you can stop right now. It cannot be done. With the rubber sheet it was easy, but when dealing with three-dimensional space we cannot picture any additional dimension. Instead we have to allow the mathematics to guide us. To use yet another analogy, we are rather in the position of a pilot trying to land an aircraft at night in a fog. He would much prefer to be able to see the layout of the runway approach for himself, but this is not possible. Instead he has to rely on his instruments and readings to guide him home. And that is how it sometimes is with fundamental physics. Explanations in terms of familiar mental images might not be possible. The explanations take a mathematical form instead.

I must confess that I have always had a hankering for this finite closed solution to the problem of the size of the universe. It is such a neat, brilliant piece of lateral thinking. It is a possibility allowed by the theory of relativity. The trouble is that the same theory that permits this kind of solution also attributes the curvature to a specific cause. It is all due to the contents of the universe. The greater the density of matter and energy in the universe the greater the curvature. One finds that there is a particular density—the **critical density**—such that if the actual density in the universe exceeds this value, the space will curve back on itself and we have a closed universe. So we tot up all the contents of the universe and see what the density is. When we add up all the matter that we can see—that making up the stars and the interstellar gas—we find it comes to no more than 5% of the critical value. But we must not stop there. On examining how the stars rotate about the centre of their galaxy we discover that they are moving much too fast to be held on course by the gravity exerted by the rest of the visible material of the galaxy. This has led to the realisation that the material we can see is but a fraction of the total. The rest is labelled **dark matter**. 'Dark' because it is matter that emits no light. Although we cannot see it, we know it has to be there. Indeed, we know how much of it there is. This we calculate from the strength of the gravitational force required to keep the stars on track as they orbit the centre of the galaxy.

Furthermore it is noticed that the 30 or so galaxies that make up the cluster to which our Milky Way Galaxy belongs (the so-called **Local Group**) are moving about too quickly to be held together by the gravitational forces exerted by the galaxies—even including the dark matter within the galaxies. This in turn is interpreted to mean that there is additional dark matter in between the galaxies.

Altogether, dark matter adds up to 25% of the critical density. What the nature of this dark matter is we do not as yet know. I suspect it is only a question of time before it is identified, but I cannot be sure.

What is the nature ≺
of dark matter?

Adding together the observable matter and the dark matter still leaves a shortfall of 70%. However, this is made up of dark energy—a property of space itself. This we shall be discussing in detail later. Thus it turns out that the actual density of the universe is exactly the critical density! A coincidence? Not really. It is due to inflation. It is built into the mechanism of inflation that it creates new matter. Most of the matter we see around us today did not, in fact, originate at the instant of the Big Bang; it was created a fraction of a second later during inflation. Moreover, the amount produced is such that the overall density should end up with exactly the critical value. The agreement between this requirement and the experimentally measured value for the mean density of the universe, provides powerful evidence in favour of inflation theory.

So what does this mean in terms of the size of the universe? Because the density is critical—and does not exceed that value—three-dimensional space does *not* curve back on itself. So the closed universe hypothesis—attractive though it might be—is dead.

This, in fact, is not the first time that a beautiful cosmological theory has died. You might have heard of the **steady state theory**. For a time, this was a rival theory to the Big Bang hypothesis. It held that the universe had no beginning and will have no end. If you were to look at a certain region of space you would see the galaxy clusters moving out from it as a result of the expansion of the universe. But new matter was continually being created, in the form of light elements, and this new matter would collect together to form new stars and galaxies, which in their turn would move out of the region. Essentially the appearance of that region of space would remain unchanged. It was in a steady state, with the loss of material through the recession of the galaxies being exactly balanced by the creation of new matter. Thus there was no need to invoke a one-off Big Bang. It was an aesthetically pleasing alternative which appealed to many cosmologists. The only trouble with it was that, over the years, the evidence for the Big Bang became so overwhelming that the steady state theory had, reluctantly, to be discarded.

And so it is with the hypothesis of the closed, finite universe. The hard evidence is against it, so it too must be set aside. Pity. It means that we are left with the answer that the universe is infinite. But what kind of answer is that? What do we actually mean by saying it is infinite?

I am always suspicious of that word 'infinite'. Ever since I heard the story of the infinite hotels. You know the one I mean? There was this rich man who built a hotel with an infinite number of bedrooms. Business was good and every room was filled. So he built another hotel alongside the first. It also had an infinite number of bedrooms. Business was *very* good and that one was also full. Then disaster struck. One of the hotels burned down in the middle of the night, so the hotelier had an infinite number of guests out in the street freezing to death. But then he had a brain wave. He told everyone in the remaining hotel to look at the number on their bedroom door, double the number, and move to that new bedroom. All those guests had a room to go to. But in doing this, all the odd numbered bedrooms became free. And as there were an infinite number of odd numbered bedrooms, all the guests out in the street from the other hotel had a room in the surviving hotel— despite the fact that the surviving hotel had originally been full! Which just goes to show how wary one has to be of that word 'infinite'. It leads me to suspect that the assertion that the universe is infinite in size might well be nothing more than a way of disguising the fact that we simply do not know how to answer the question: How big is the universe?

Is the universe ≺
infinite in size, and if
so, what exactly does
that mean?

6

Extraterrestrial life

When thinking about the universe, one cannot help wondering whether there is life out there. As far as the planets of our Solar System are concerned, there might be some forms of very primitive life. We should have the answer to that quite soon as a result of planned space probes. But not intelligent life. No, if ET exists it must be on a planet going round some other star.

And there is no shortage of them. One of the fastest developments in modern-day astronomy is the identification of these planets. They can be located by examining how certain stars appear to wobble—they exhibit a circular motion as they orbit the centre of mass between the star and the unseen planet. A second method is to study the dimming of the light from the star as the planet passes between it and ourselves, partially blocking out the stellar light. A third method, recently developed, blocks out completely the light from the star thus enabling the much dimmer light being reflected from the planet to be seen. This last method holds out the possibility of being able to detect seasonal changes in colour of the planet, which might indicate the presence of vegetation of some kind, and hence the existence of at least some form of life. It might also reveal by spectral analysis the chemical composition of the planet's atmosphere. One would be looking especially for oxygen, water, ozone, and carbon dioxide as indicators of life. At the time of writing over 300 of these planets have been found, but the tally increases weekly. Most of these **exoplanets**, as they are called, will not be habitable—like most of the planets in the Solar System. Given that there are 100 billion galaxies in the observable universe, with an average of 100 billion stars per galaxy, the law of averages would say there must be many, many Earth-like planets—planets where intelligent life *could* flourish.

Which is not to say that it *has* flourished. It is conceivable that the evolution of intelligent life here on Earth might have involved a massively improbable set of circumstances—so improbable that it is unlikely to have occurred a second time anywhere else in the cosmos. I myself find such a scenario unappealing. Ever since we humans abandoned the idea of the Earth being the centre of the cosmos, I reckon we would do well to discard any thoughts that we are unique. That being so, how are we to see ourselves in comparison with ET? Where, for example, do we rank in the pecking order of intelligent beings in the cosmos?

Over the past 3 million years of human evolution the brain has increased in size compared to body size by a factor of more than three. Our human species, or as some would have it, our subspecies—*Homo sapiens sapiens*—is young: little more than 100,000 years old. These periods of time are exceedingly short on the evolutionary timescale of some 4 billion years. In considering how evolution might have proceeded on other planets, the process there could have got out of step with what has happened here on Earth by only a little to give rise to a situation where ET reached our level of intellectual competence a million, ten million, or a hundred million years ago. That being so, are we to conclude that by now they must have evolved to a point where they far exceed human capabilities?

Here we need to be careful. Certainly a steady increase in intellectual capacity has been the distinguishing characteristic of human evolution so far, but that does not automatically mean that the same process will continue indefinitely. The reason why in the past the general intelligence level of the population has increased is that there has been survival value in being intelligent. Under circumstances of a shortage of food and shelter, and a need to resist predators, those fortunate enough to be endowed with greater intelligence had an enhanced likelihood of being able to survive to a point where they could mate and have offspring that inherited their greater intelligence. The less intelligent fell by the wayside before they were able to pass on their less favourable genes. Hence the succeeding generation, on average, had a higher level of intelligence than the previous one, the process then being set to repeat itself.

But the situation has now changed. There are social factors to be taken into account. In our civilisation it is probably true to say that, on average, the more intelligent we are, the better chance we have of securing a well-paid job. But we know that the higher our standard of living, the *fewer* children we are likely to have. This militates against selection for intelligence. Indeed, it might well be that the process has now gone into reverse and that, at a deep level, the average intelligence of the population is on the decline—a development that at present is being temporarily masked by exposure to better education. Thus it could well be that ET reaches a point where selection for intelligence no longer applies to them, and from then on they progress no further.

That is one scenario. But there is another. The **human genome project** was set up to unravel human DNA codes. With a greater understanding as to which codes correspond to which human features comes the possibility of deliberate genetic modifications to the species. The discovery of which codes correspond to genetic diseases such as sickle cell anaemia holds out the promise of being able to eliminate such defects from the gene pool. Likewise, if one finds genes associated with what are regarded as beneficial characteristics, such as greater intelligence, then there opens up the possibility of enhancing such characteristics by genetic engineering. At the present time there is a reluctance to go down the path of 'designer babies'; it smacks too much of Aldous Huxley's *Brave New World*. But that is not to say the human race will always have such inhibitions. Or that ET has them. ET having reached our stage might well have embraced such possibilities and gone in for enhancing intelligence. So, rather than stagnating, it could well be that ET has launched itself into an unparalleled explosion of evolution—what we might call **directed evolution**, in contrast to the earlier evolution by natural selection. The former would be based on deliberate choice, the latter having been dependent on chance mutations to the genes. In that case it could well be that ET would regard humans in much the same light as we regard our more primitive prehuman ancestors.

There is yet another scenario to consider. Once one gets to our stage of evolution one discovers nuclear power—the power to destroy oneself. Though at the present time the threat of a nuclear holocaust has seemingly receded—certainly compared to the days of the Cold War and the 1962 Cuban missile crisis—it has not gone away, and will never go away. Can anyone seriously have confidence that over a timescale of, say, a million years, no deranged dictator will come to power in a country with a nuclear capability, or that no nuclear war will be started by some mistake like a computer malfunction, or there will not be some dreadful monster-sized Chernobyl accident? I am not saying that any of these events is imminent in the next few years, 100 years, or even 1,000 years. But such spans of time are nothing compared with those appropriate to evolutionary change—which is what we are considering here. Who knows, it might be an invariably repeated pattern of events throughout the cosmos that species evolve to roughly our stage, they discover nuclear power, and a short while later, before there can be any further significant development, they annihilate themselves.

These are fascinating speculations that involve not just the evolution of intelligence but also individual choices and societal factors. We would love to know whether any of them have any truth in them. Admittedly questions associated with ET are not in the same class of ultimate questions such as: Why is there a universe? Why are there laws of nature? and so on. Nevertheless, establishing the true status of humans in the cosmos is undoubtedly a fundamental question of sorts. How could we set about answering it?

Go and visit them seems to be the obvious answer—as is done in countless science fiction adventure stories—but this is hardly a realistic possibility. Travelling at speeds typical of today's space probes it would take close on 100,000 years to reach even the nearest star. So, space travel is unlikely to be the answer.

The other approach is to search the skies for any signals indicating that extraterrestrial intelligences are trying to communicate with us. Such searches have been carried out for many years by the SETI programme—the Search for Extraterrestrial Intelligence—but so far without success. The search will carry on, but there is no guarantee of success. The situation is not under our control. It is up to ET (if they exist) to take the initiative and contact us. Perhaps ET declines to do so. Perhaps ET does not have the kind of transmitters that could send a signal detectable over so great a distance. Thus the question might remain unanswerable—not for any profound metaphysical reason—but just for practical reasons.

> *Is there extraterrestrial life, and if so, how do we humans stand in comparison as regards intellectual capacity?*

7

The nature of space

What actually is space—space itself? By this I do not necessarily mean outer space—the cosmos. I am talking of space in general—the kind of space in the room where you are sitting. As we previously noted, the obvious answer seems to be 'nothing'. If there are no objects in a particular region of space, we speak of it as being empty space—which seems to imply nothingness. And yet that is not how the modern physicist views it.

We have already had a hint of this. Recall how, earlier, in describing the nature of the Big Bang, I said that it was not like other explosions. The galaxy clusters were not receding from us by moving through space; it was space itself that was expanding and carrying the clusters along on a tide of expanding space. That hardly sounded like nothing! And again, when considering the size of the universe we were prepared to contemplate space possibly being curved back on itself. That seems to suggest that space is a something—an entity that can be curved. We discovered in fact that space, considered overall, is not curved. It is flat—this being tied to the fact that the average density in the universe has the critical value—as required by inflation theory. So, is that the end of the story? Space is not something that is curved? We must not be hasty. All we have spoken about so far is the overall curvature of space across the universe. We have yet to examine what happens on a more local scale.

Take for example the space surrounding the Sun. The Earth moves in orbit about the Sun. Why? Because of the force of gravity exerted on it by the Sun. It is that force which keeps the Earth from flying off into outer space. Everyone learns that at school. In the same way, the space station moves in orbit about the Earth because of the force exerted on it by the Earth.

That, at least, is how one conventionally describes the situation. However, there is something very odd about the gravitational force. We can see that by considering an astronaut stepping outside the space station and floating in orbit alongside it. Both the astronaut and the station are essentially travelling in the same path. But the astronaut is much lighter than the spacecraft, so it takes less force to keep her on course than it does the craft. Thus gravity has to pull less strongly on her than on the craft. Which it does. Gravity somehow seems to know exactly how hard to pull on objects to make their motions the same. How does it know? And in any case, *why* would it want to keep both the craft and the astronaut on the same path?

The answer from relativity theory is to do away entirely with the concept of gravitational forces. Instead it holds that a better way of looking at things is to say that what the Earth is doing is not exerting a force on the objects orbiting it, but instead it is curving, or warping, the surrounding space. As a result of this distortion of space, anything passing through the affected region no longer travels in a straight line. Instead it follows a curved trajectory. The spacecraft and the astronaut follow the same path because that is the *natural* path for *any* object to follow if they start off from the same point with the same velocity. The natural path is not a straight line—a straight line requiring a force to convert it into an orbit. It is as though the Earth has caused a dimple in space—much as a ball bearing would if placed on a stretched rubber sheet (Figure 2). A second ball rolling across the sheet—say, a light table tennis ball—would follow a curved path because of the curvature of the sheet.

It might even find itself captured into an 'orbit' repeatedly circling the ball bearing. So, the Earth causes a dimple in three-dimensional space. And the Sun an even more pronounced dimple such that the Earth finds itself describing an orbit around it. As for the Sun, that too describes an orbit—one that circles around the centre of the Milky Way Galaxy—this being due to an enormous dimple caused by the 100 billion other stars in the galaxy.

This way of viewing gravity—which we owe to Einstein's general theory of relativity—comes up with more accurate

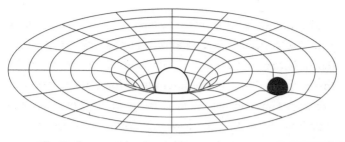

FIG 2 The Earth causes the surrounding space to curve, it being this curvature that is responsible for spacecraft and space walkers following curved orbits.

descriptions of nature than Newton's ideas based on an inverse square law of gravitational forces—and so is to be preferred. But there is a price to be paid. We have to take on board the somewhat counterintuitive idea that space has no longer to be regarded as nothingness; it is a something, a mysterious something that can be curved.

What more might be said about it? Quantum theory—the theory of the very small, that I mentioned earlier—requires that so-called empty space is not empty at all. Instead it is a seething crowd of subatomic particles popping into and out of existence. It is a manifestation of the basic uncertainty that afflicts everything going on at the subatomic level.

But, you might ask, how can particles, which were not there before, suddenly come into existence? What about the commonsense idea you might have learned at school that 'matter can neither be created nor destroyed'? The simple answer is that this statement is not true. According to relativity theory, matter is just another form of energy—much like the others:

kinetic energy (for example, the energy of a car in motion), potential energy (like that of a brick that is about to be released from a height), the energy of the Sun's rays, heat energy, etc. Energy locked up in the form of matter can only be released under certain special circumstances—the fusion of atomic nuclei in stars, for instance. Fusion causes a proportion of the energy contained in the hydrogen and helium nuclei to be converted into other forms such as light and heat. And just as matter can be converted into other energy forms, so energy can be used to produce matter. This is what happens in the great particle accelerators that we were talking about earlier. The kinetic energy of the accelerated subatomic particle, on crashing into another particle, leads to the production of new particles—new matter.

This is a possibility encapsulated in Einstein's famous equation:

$$E = mc^2$$

This states that energy, E, and mass, m, are equivalent. Kinetic energy of motion—the energy carried by the bombarding particle—can be transformed into a different kind of energy—a locked-up form that manifests itself as matter.

Now that is all very well, you might want to argue. When creating new particles in accelerators, we are not starting out with nothing; we start with energy. All we are doing is converting pre-existing energy from one form into another. In empty space, on the other hand, where is the energy to start with? There is a law called the law of conservation of energy. It maintains that whatever energy you have at the beginning (in all its forms) will be the amount of energy you will have at the end (even though some may have changed from one form to another). So if, as seems to be the case with empty space, there is no energy to begin with, you're stuck. Without energy you can do nothing. That's the end of it.

Except it isn't! This is where Heisenberg's uncertainty principle comes in. While it agrees that the law of conservation of energy must hold in the long term—averaged over a period of time—there can still be minute fluctuations over shorter times. A physical system can 'borrow' energy, provided it pays it back within the brief time specified by the principle. It is similar to a short-term loan to cover a firm's temporary cash flow problem, rather than an extended mortgage for a house purchase. As we shall see later, when we come to consider quantum theory in more depth, there is a simple equation:

$$\Delta E \times \Delta t \approx h/2\pi$$

where ΔE is the energy borrowed, Δt is the payback time, and h is a small constant called **Planck's constant**. From this equation we can deduce, for example, that in order to borrow twice the energy, it has to be paid back in half the time. The reason these fluctuations are not obvious in normal everyday life is that the constant on the right-hand side of the equation is tiny. Violent energy fluctuations, were they to occur, would be too fleeting to notice.

But on the subatomic scale, the energy fluctuations can be significant. All of atomic and subatomic physics must take account of quantum uncertainties. And in the present context, when thinking of the nature of space, we must allow for subatomic particles to be popping into and out of existence thanks to this fundamental uncertainty. We do not actually *see* these particles; they disappear again much too quickly. For this reason, they are called **virtual particles**. They are expected to be formed in pairs rather than singly: particle—antiparticle pairs. Antiparticles have the same mass as their particle, but opposite values for other properties, such as, for example, electric charge. Thus, an electron has negative electric charge, whereas the antielectron—called a **positron**—has positive charge. The proton—one of the constituents of atomic nuclei—has positive electric charge, whereas the antiproton has negative charge. The creation of a virtual particle—antiparticle pair will therefore not violate other physical laws such as the law of conservation of electric charge.

It is in this way that the physicist has further reason to consider so-called empty space to be not empty at all. We have already seen how it is something that can push galaxy clusters apart and can, in the presence of matter, be curved. Now we see that it is a seething mass of unseen virtual particles popping into and out of existence.

All this activity is expected to give rise to an overall average energy. (The average energy density certainly cannot be zero if, for some of the time, there are particles around—albeit virtual particles.) This energy—the energy character-istic of empty space, i.e. the vacuum—is the so-called **dark**

energy that we mentioned in connection with totting up the average density of the universe. If you recall, it accounted for 70% of the total, so it is not something that can be ignored. Not that it has so far proved possible to calculate how much dark energy there should be. Indeed, the calculated amount comes out 120 orders of magnitude greater than the actual value—which is very embarrassing!

> *How are we to account for the observed value of the dark energy?*

Though we never actually see the virtual activity that is causing the dark energy, the dark energy is expected to be there. Indeed, Einstein himself was aware, from his equations of general relativity, that empty space could have an energy density itself—an energy density that would be the same everywhere and would have the same value for all observers regardless of their motion. Though the energy cannot be seen directly in the form of virtual particles, it is expected to produce detectable effects. Such effects depend on the sign of the energy. A positive energy density would produce an expansion—moreover, one that was accelerating.

Until recently it was thought that, although the universe is expanding, with the galaxy clusters separating to greater and greater distances, they will be slowing down because of the mutual gravitational attraction between them. We earlier considered the possibility that this attraction might one day bring everything in the universe back together again in a Big Crunch. But in 1998 it was discovered that the distant galaxy clusters, far from slowing down, were actually accelerating away from us. It is now believed that in the early stages of expansion (after the brief period of inflation) there probably was indeed a slow-down due to gravity. But operating against this attraction was the repulsion of dark energy. With the expansion of the universe being due to an expansion of space, there was ever more space being created, and with it more and more dark energy to go with the new space. It is the enhanced repulsion due to the increasing amount of dark energy that has now led to us passing over into a phase of the universe's expansion where dark energy has at last won the day in its battle against the gravitational attraction, and the clusters are no longer slowing down but are accelerating away from us. And all this is due to the action of so-called empty space! This in itself raises an interesting question as to where the universe gets all this new energy from. It appears that even our hallowed law of conservation of energy, which has proved so helpful in understanding how the component parts of the universe interact with each other, does not apply in this overall regard.

Before leaving this topic you might like to hear of two further speculations concerning dark energy. One is the possibility that the density of dark energy might not be a constant characteristic of space but might increase with time. If so, the repulsion, which at present can be seen only over the kind of distances separating galaxy clusters, should become progressively more noticeable over smaller and smaller distances. First galaxy clusters will be torn apart, then the galaxies themselves will disintegrate, then the solar systems, the stars, atoms, and finally the nuclei. This is called, somewhat tongue-in-cheek, the **Big Rip**.

Secondly, it has been suggested that there might be some sort of connection between the mechanism responsible for the repulsion of the galaxy clusters we see happening today, and the repulsion that was manifest at the time of inflation. This remains an open question.

➤ *Does the density of dark energy remain constant with time?*

All very mysterious. And there is yet more to learn about the nature of the vacuum. Just now I mentioned how particles had antiparticles. The existence of antiparticles was actually predicted by Paul Dirac long before they were experimentally found. (Here I am talking about *real* antiparticles such as are produced in particle accelerators, not the virtual ones.) How did he do it?

➤ *Is there a connection between today's repulsion of the galaxy clusters and the period of inflation?*

He was engaged in seeking an expression for the energy of an electron, taking into account the new insights that had become available through Einstein's theory of relativity. The details do not concern us. Suffice to know that the last step in the derivation involved taking a square root. Now, as is well known, two solutions arise whenever one takes a square root—a positive solution and a negative solution. For example, the square root of 4 is either $+2$ or -2. Thus Dirac found that the mathematical solution to his problem yielded a positive value for the energy of the electron and a negative one. The first solution correctly described the behaviour of the electron. But what about the negative solution? In this sort of situation it is customary to discard the negative value as being 'non-physical'—a quirk of the mathematics having no practical significance. After all, if an electron did have a negative energy it would mean that its locked up energy, i.e. its mass, would be negative. That would imply a particle such that when you pushed on it, it would come towards you, and when you pulled on it, it would move away! Clearly we know of no such behaviour. So the sensible thing would simply have been to ignore the negative solution as just a mathematical oddity.

Dirac, however, thought differently. By an astonishing piece of lateral thinking, he suggested that the reason we never saw negative mass electrons was not because they did not exist. On the contrary, we did not see them because there were so many of them! They were everywhere—literally *everywhere*. They filled up the whole of space—even the space between the nucleus and the electrons of an atom. Negative energy electrons formed a continuum—a perfect continuum—and a feature of a perfect continuum is that it is undetectable; it cannot be observed. To be able to observe something—a chair, say—it must be characterised by being at a particular location in space. You need to be able to point to it and say, 'I am talking about that; I am not talking about anything else in the room.' But in the case of the continuum of negative energy electrons, where does one point? Everywhere—and nowhere in particular.

As an analogy, think of the air in the room where you are sitting. It approximates to a continuum of sorts. It is both around you, and also in your lungs. Sitting perfectly still and not breathing (only for a short time, of course) you are unaware of the air. To become conscious of it you inflate your lungs and blow it out. Or you wave a sheet of paper about. But note that in doing so, you are disturbing the continuum. The act of breathing changes the density of the air in your lungs compared to that outside. In the same way, there is a build-up of density in front of the moving sheet of paper, and a diminishing of density to the rear. Thus the air is no longer a uniform continuum. The continuum that Dirac envisaged, on the other hand, was perfect. Your lungs cannot compress or expand it; a moving sheet of paper just passes through it experiencing no resistance; the continuum passes through the sheet—through its atoms—undisturbed and undetected.

Dirac did not let this difficulty put him off. He hit on a way that might disturb the continuum. He envisaged a particle, or a packet of light energy (called a **photon**) moving through space and hitting one of these unseen negative energy electrons. In doing so, it could be that the impact was so great, and the consequent energy transfer so significant, that the negative energy electron acquires sufficient energy to convert its negative mass into a normal positive mass. This being the case, the electron would no longer be part of the continuum of negative energy electrons, and would thus suddenly become visible; it would appear to have popped into existence from nowhere. Not only that, it would leave behind a 'hole' in the continuum. What would that look like? It would be a loss of a negative mass—which is equivalent to a gain of positive mass. So the 'hole' would appear to be a particle with the normal mass of an electron. What else? The original negative mass electron would have had a negative charge like any other electron. But now we have the loss of a negative charge—which is equivalent to the gaining of a positive charge. So the 'hole' would show up as a positively charged particle with the same mass as the electron—the antielectron, or positron. In summary, such a collision would give rise to a pair of particles—an electron and a positron. And indeed this is exactly what is observed. Particle—antiparticle pair production is now a well-established phenomenon.

This then is another indication that empty space is not to be regarded as simply nothing. It is packed with negative mass electrons—and also with the negative mass versions of many other fundamental particles described by Dirac's equation, such as the proton and the neutron.

Or is it? Although it is true that the negative energy continuum was the route by which Dirac came to his prediction of the existence of antiparticles and how they might be produced—a discovery for which he was awarded the Nobel Prize—the question arose as to whether this actually was a true description of reality. I recall a conference, held in Jerusalem in 1979 to mark the centenary of the birth of Einstein, at which Dirac spoke. During one of the coffee breaks I overheard another Nobel Prize winner exclaiming, 'I have just spoken to Dirac. Guess what. He told me he believes these negative energy particles actually exist. That's right; that's what he said. They actually *exist*! How can anyone believe *that*?!' Opinion is divided. Some physicists will have nothing to do with the continuum idea. OK, Dirac happened on the idea of antiparticles by what some regard as a quirky line of reasoning, but just because he got the right answer— namely, particles such as electrons, protons and neutrons have antiparticles—it does not necessarily follow that the route he took was the correct way of looking at things.

While on the subject of antimatter we ought to mention a problem that has yet to find a satisfactory resolution: Why do we find in nature more matter than antimatter? One might imagine, for instance that in the highly energetic conditions of the Big Bang it would be just as easy to create antiparticles as particles. In attempting to answer this question, we note from high energy experiments carried out in particle accelerators, the interactions of particles and their antiparticles are known in certain respects to be very slightly different from each other. Such differences, in the early stages of the development of the universe, could have led to a slight preponderance of ordinary matter over antimatter.

Couple this with the fact that many particles would over this period be meeting up with their antiparticle and annihilating each other (i.e. the particle pair converts back to energy) and we could have the situation where all the remaining antimatter gets destroyed, leaving just 'the slight preponderance of matter' as the sole remnant in today's universe. Whether the slight difference between the interactions of matter and antimatter is sufficient to produce such a result is unclear. In any case, we still seem to be left with the question as to why the difference in the interactions should favour matter over antimatter, rather than the other way round.

Why is there more matter than antimatter? ≺

We said that just because Dirac happened to come across the idea of antimatter in the way that he did, does not necessarily mean that his was the most helpful way of viewing things. The same sort of caveat might apply to the other insights we have had into the nature of space—the notion that it can be curved, for instance. All we actually know for certain is that the mathematical equations of relativity provide a more accurate description of motion under gravity than Newton's earlier inverse square law. It is *as if* space were something that could be curved. But is it? Would it not be preferable just to stick to the mathematical description rather than worrying our heads over some supposed physical explanation of what is actually going on? In the same way, is it truly the case that space is a seething mass of virtual particles giving rise to dark energy, or would it be preferable once again to stick with what we actually observe—galaxy clusters accelerating away from us.

➤ *How are we to understand the true nature of space?*

One further thought about space. We are accustomed to thinking of it as being divisible into ever smaller distances: kilometres, metres, millimetres, nanometres, and so on. But does that go on indefinitely? Is there no limit to how small a division of space can go? Or do we eventually get to a basic unit of space that is no longer divisible?

At present we have two great physical theories: quantum theory and general relativity. The eventual aim is to reconcile these two into a combined theory of quantum gravity. Such a theory is bound to depend very heavily on three fundamental constants: (i) the gravitational constant, G, governing the strength of gravity in both Newton's theory of gravity and Einstein's general theory of relativity; (ii) Planck's constant, h, governing the size of quantum fluctuations, and (iii) the speed of light, c, which figures prominently in both relativistic and quantum physics. From these three constants, the German physicist Max Planck, one of the founders of quantum theory, found that there was a unique way of using them to define a quantity with the dimensions of length:

$$l_P = (\hbar G/2\pi c^3)^{1/2} \approx 1.6 \times 10^{-35} \text{ metres}$$

This quantity is named the **Planck length**. It is very small—about 10^{-20} times the size of the proton. It is often regarded as in some sense the 'natural' unit of distance. What one can say is that any attempt to unite quantum theory with relativity must involve these three constants, and hence any expressions of a distance arising out of the combined theory will involve this unit of length, possibly multiplied by some unimportant numerical factor such as 2π. According to some theories of quantum gravity, this is expected to set the scale where quantum fluctuations become so pronounced that the very structure of space itself breaks up and it becomes a discrete kind of foam—thus setting a limit on the smallest distance that can still have properties recognisable as those of space. However, there are other attempts at formulating a theory of quantum gravity that do not point to such a conclusion. The position is unclear, and could remain so.

Is space infinitesimally divisible? ≺

The same kind of reasoning applies to time. We divide up time into ever smaller intervals: years, months, weeks, days, hours, minutes, seconds, milliseconds, and so on. But is time infinitesimally divisible, or is there a smallest unit of time? Once again we look to our three fundamental constants. We find that we can devise an expression with the dimensions of time:

$$t_P = (hG/2\pi c^5)^{1/2} \approx 5.3 \times 10^{-44} \text{ seconds}$$

This is named the **Planck time**. It is the time it would take light to travel a distance equal to the Planck length. But, so it might be argued, if a distance in space equal to the Planck length might not be meaningful, perhaps this also implies that a time interval equal to the Planck time might be the limiting interval that can be defined.

➤ *Is time infinitesimally divisible?*

The trouble with the Planck length and Planck time, of course, is that both are almost inconceivably small compared to any spatial distance or time interval that has ever been measured, or is ever likely to be measured. The smallest distance that has yet been probed experimentally is 10^{-18} metres—a factor 10^{17} bigger than the Planck length, and the shortest time interval is 10^{-16} seconds—a factor 10^{27} bigger than the Planck time. This in turn raises the uncomfortable thought that perhaps this is another example of the secrets of nature remaining elusive for practical reasons. Crucial features of the world might involve scales that are in no way matched to what humans can investigate experimentally. Acceptance of the idea that these measures of space and time really do have some physical significance will, therefore, have to rest on arguments based on more indirect observations.

8

Space in relation to time

aving discovered how complicated mere empty space (or 'nothingness') can be, it is with some trepidation that we approach the next topic: time. As you will recall from my opening remarks, it was largely Einstein's ideas about time and space that originally fired my interest in physics.

There is an intimate relationship between space and time. It all begins with the speed of light—customarily denoted by the symbol, c, and having a value of about 300,000 kilometres per second. Strictly speaking, this is its speed in a vacuum, but it is almost the same as in air (it is slower when passing through glass or water). Light is made up of electric and magnetic vibrations moving through space. It turns out that, knowing the measured constants governing the strength of electric and magnetic forces, it is possible to calculate how fast electromagnetic disturbances such as light, should move. This emerges out of James Clerk Maxwell's laws of electromagnetism, formulated in 1864.

Physical laws apply equally to all observers in uniform steady motion. This has been known since the time of Galileo and is called **the principle of relativity**. If, for example, we measure the strength of electric and magnetic forces while flying in an aircraft, we should find that the same values apply when the identical experiment is carried out on the ground. Because Maxwell's laws hold equally in the aircraft as on the ground, we can immediately conclude that, on feeding the identical values for the strengths of the electric and magnetic forces into the relevant formula, we will arrive at the same value for the speed of light, 300,000 kilometres per second, whether or not we are moving. In other words, all observers in uniform relative motion, when they measure the speed of light, should find the same value.

In practice, this implies that if I am standing by the side of the road, and you are driving a car in my direction with your headlights on, then I will find that the speed of the light coming towards me has the value, c. The fact that the source of the light is moving does not affect the speed of the light emitted. In this respect, it is rather like the waves caused by a boat. The speed of a boat has no effect on the speed of the waves it is making.

How does the situation look to you, the driver of the car? At first one might think that, if light really is behaving somewhat like waves caused by a boat, then to you aboard the moving vehicle the waves ought to be moving ahead

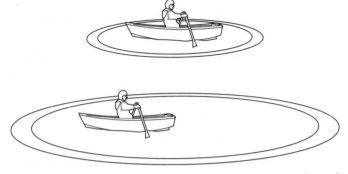

FIG 3 Water waves move away from the boat more slowly in the forward direction than they do to the rear.

more slowly. After all, in a boat the waves generated move ahead of the boat with a relative speed which is the normal wave speed minus the speed of the pursuing boat. In a similar way, the waves that are being left behind to the rear should move away at a speed equal to the normal wave speed plus the speed of the boat (Figure 3).

Thus one might expect that for you in the car, the light from the headlights at the front should move away from you more slowly than the light emitted by the tail lights to the rear. But that cannot be so. Maxwell's laws apply equally to you, the moving driver, as they did to me, standing by the roadside. This in turn means that, as far as you are concerned, the light from both the headlights and the tail lights will move away from the car at speed c.

And this is indeed what you would find in practice. Experimental measurements of the speed of light always yield the exact same value regardless of whether the source of light or the observer are considered to be in motion. The speed of light is a constant. Although, at first, it does not seem to make sense, it is nevertheless an experimental fact. It has to be accepted. Countless experiments have confirmed the conclusion.

Clearly there is something wrong with the way we customarily handle velocities. If the light from the headlights comes towards me at the roadside at speed, c, and the car is coming towards me at speed, v, then 'common sense' says that the speed of the light relative to the car must be $(c - v)$. But according to you, the speed relative to your car is not $(c - v)$, but c. In the same way, you seeing the light moving away from your car at speed c, and seeing me coming towards you at a relative speed v, will conclude that I ought to see the light approaching me at a speed $(c + v)$. But as already noted, I do not; the speed of light relative to me is c. So, our understanding of velocity is faulty.

But how? A velocity is nothing more than a distance travelled divided by the time taken. It follows that if there is something wrong with the notion of velocity, then there must be something wrong with the underpinning notion of distance, or of time—or it might be that our understanding of both distance *and* time needs to be revised.

These considerations were the starting point for Einstein's **special theory of relativity**, published in 1905. It is said to be 'special' in the sense that it deals only with the effects of uniform motion on space and time. His later **general theory of relativity** deals not only with this topic but also with the

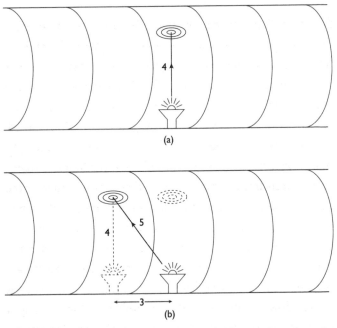

(a)

(b)

FIG 4 The passage of a light pulse sent from the floor to the ceiling of a moving spacecraft, (a) as observed by the astronaut, and (b) as observed by the mission controller.

additional effects of gravity—it being the general theory that led us to the idea we considered earlier of curved space.

It turns out that our conventional ideas about *both* space and time need to be overhauled. We can see this in the following way: Suppose we imagine a spacecraft flying at high speed on a journey from the Earth to a distant planet. The astronaut places a light source on the floor of the cabin and observes it sending a light pulse vertically upwards to hit the ceiling (Figure 4a). She uses a stop clock to time its passage from floor to ceiling. The mission controller, observing what she is doing, also times how long it takes for the pulse of light to go from floor to ceiling. According to him, the pulse has further to travel because the craft moves forward in the time it takes to complete its journey. It has to follow a slanting path (Figure 4b) rather than the shorter vertical path. As we have noted, both observers use the same value for the speed of the light pulse. The controller, seeing the pulse travel a longer distance at this speed, finds on his stop clock that it takes longer for the pulse to reach the ceiling than it does according to the astronaut. As far as the controller is concerned the astronaut's clock, with its lower reading, must be going slower than his. And not just her clock. It is the nature of time itself we are talking about. *Everything* happening within the spacecraft is slowed down according to the controller: the functioning of the craft's electronics, the astronaut's breathing and pulse rate, her aging processes, her brain processes and hence her thinking. It is because her thinking is slowed down, in exactly the same ratio as all other processes within the craft, that everything happening in the craft appears perfectly normal to her. Life carries on as usual. It is only to the mission controller that her time has been affected by the motion. The phenomenon is called **time dilation**. As an illustration, at a speed of about 0.9c, everything is slowed down by a factor of about a half.

But one might think that the astronaut is bound to know something is wrong with her time when she finds that she has arrived at her destination too soon—in half the time scheduled. After all, both astronaut and controller are agreed

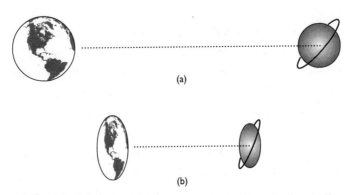

FIG 5 The Earth-to-planet distance (a) according to the mission controller, and (b) according to the astronaut.

as to their relative speed, and the distance between the Earth and the planet is known. Except that this is not so. Not the distance. Here we come across a second effect of special relativity: **length contraction**. Distances between objects moving relative to the observer—in this case the distance between the Earth and the planet as they are seen to glide past the craft—are contracted in the direction of motion (distances at right angles to the motion being unaffected). At a speed 0.9c, the Earth–planet distance, according to the astronaut, is half of what it is according to the mission controller (Figure 5). She is perfectly happy with the short time her journey took. The journey took half the time because she has gone only half the distance!

All distances moving relative to the astronaut are contracted in the direction of motion. According to her, both the Earth and the planet themselves are distorted; they are squashed in the direction in which they appear to be moving past her.

So what we find is that each observer—the mission controller and the astronaut—has his or her own internally consistent set of readings for the distance travelled and the time taken. The trouble is, when they compare their readings, they disagree with each other. And there is no way to decide who is 'right' and who is 'wrong'.

But, you might wish to argue, surely we must give greater weight to the controller. After all, he is the one who is stationary. It is the astronaut's distances and times that are questionable because of the way they have been affected by the craft's motion. Not so. All motion is relative. (*Relativity* theory does not get its name for nothing!) The craft moves relative to the Earth, but the Earth is moving relative to the craft. There is in fact nothing stationary about the Earth. It is orbiting the Sun, the Sun is orbiting the centre of the Milky Way Galaxy, and the galaxy is moving about within the cluster of galaxies to which it belongs—the Local Group.

Indeed, the weirdness of relativity goes even deeper. We see this when we consider what the astronaut regards is happening on Earth. As far as she is concerned, she is the one with the stationary viewpoint, and it is the Earth that is seen to be moving away from her at $0.9c$. That being so, using exactly the same line of argument as the controller used when observing what was happening in the craft, the astronaut concludes that time is running at half the normal rate on Earth—this being due to the Earth's motion relative to her.

This, of course, is the exact opposite of what most people might have expected. If the controller was right in saying

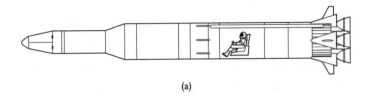

(a)

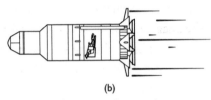

(b)

FIG 6 The shape of the spacecraft (a) according to the astronaut, and (b) according to the mission controller.

that her time is half of his, then surely she ought to conclude that his time is twice hers. But if that had been the case, one would have been able immediately to conclude that it was she who was *really* moving and not the controller. But that violates the idea that motion is *relative*. So, whatever type of conclusion the controller comes to regarding the astronaut's surroundings, the astronaut must come to the same sort of conclusion regarding the controller's surroundings.

And this applies as much to length contraction as it does to time dilation. If she regards the Earth–planet distance to be squashed down to half its normal length as it goes by her, then he will conclude that her spacecraft is squashed to half its normal length as it goes past him (see Figure 6). And not just the craft. Her body will be squashed in the direction of motion. Not that he expects her to feel anything is amiss. The atoms that make up her body are squashed to half their normal size and so do not require as much space to fit in. Neither is the controller surprised that she does not even *see* that anything is different from normal. This is so because, according to him, the retina at the back of her eye, when she looks at her contracted craft, is reduced to half its normal size. The image of the (squashed up) craft formed at the back of the eye still fills up the same proportion of the (squashed up) retina as it did before. And thus the signals going to the brain remain normal.

In summary, both observers reckon that their own immediate surroundings are perfectly normal. Each of them attributes the unfamiliar behaviour—time dilation and length contraction—to the physical set-up that is moving relative to them.

But you might wonder, who in the end is *right*? What is the *true* length of the craft? What is the *true* duration of the journey? These are questions that have no answer. Ascribing a value to a distance or a time interval can only be done in the context of a specific observer—one whose motion relative to the observed phenomenon has been specified. In the case we have been considering, there are time and distance measurements according to the astronaut, and time and distance measurements according to the controller. And that is all that can be said. Einstein's special theory of relativity has now been a cornerstone of standard physics for over a century.

Having said that, I should perhaps mention in parentheses that there has been a recent suggestion that special relativity as formulated by Einstein might in fact not be the final word on the subject. In 2002, Giovanni Amelino-Camelia proposed a variant called **Doubly Special Relativity**. In it he suggests that, if the Planck length and Planck time have the kind of significance that some physicists suggest they have, namely that they are the smallest measures of space and time, then all observers ought to agree on their unique values as derived from the constants, G, h, and c. But suppose that an observer were to be moving relative to some object that had the dimension of one Planck length (a 'bubble of spatial foam' for instance). Would she see this to be length contracted? If so, how would she reconcile that with the theoretical value she can derive from the three constants? To get round this problem, it was suggested that lengths that approach the Planck length might not undergo the usual contraction. Also, times as small as the Planck time would not get dilated. Observers should therefore agree not only on the value of the speed of light, but also on the Planck measures. However, such a proposal is for the time being at least, just speculation.

So, let us return to normal, accepted, special relativity. All this talk of observers having their own length and time measurements sounds very messy and confusing. But all is not lost. There *is* actually something that all observers do agree about. I like to describe it with the following analogy:

I hold up a pencil in a room full of people. The appearance of the pencil is different for everyone. What it looks like will

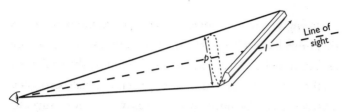

FIG 7 The apparent length of a pencil varies according to the angle at which it is viewed.

depend on where one happens to be sitting relative to the pencil. Some will see a long pencil because they are seeing it broadside-on, while others will be getting a foreshortened end-on view. In other words, the flat two-dimensional images of the pencil formed on the retinas of the viewers' eyes will be different. Does that worry us—the fact that they don't agree as to what the pencil looks like? No. We are all aware that a two-dimensional image is but a projection at right angles to the line of sight to the object—an object that actually exists in three dimensions (Figure 7). On taking into account how much of the pencil lies *along* the line of sight, all observers are agreed as to what the length of the pencil is in three dimensions. Those sitting broadside-on add very little to what they see in projection because the pencil extends only a little way along their line of sight; those sitting end-on have to add in a large contribution for the extension along their line of sight. So everyone is happy. There is no mystery about the different appearances—they were but projections; they were not the 'real thing'.

We now apply this same reasoning to the case of the spacecraft journey between the Earth and the distant planet.

We have seen how the two observers, the astronaut and the controller, have different ideas as to the distance travelled by the craft and the time taken. Suppose these too were but 'appearances'—projections of some sort. But of what? The modern view of space and time is that they are not separate entities—a three-dimensional space and a one-dimensional time. Instead, they together constitute a four-dimensional **spacetime**. The time axis becomes the fourth dimension, and is as indissolubly welded to the spatial axes as the three spatial axes are welded to each other. This, incidentally is the reason why we earlier concluded that if the instant of the Big Bang saw the coming into existence of space, which subsequently continued to expand, then it must also have seen the coming into existence of time.

The idea that space and time are so intimately related at first sounds unlikely, to put it mildly! After all, we perceive and measure spatial distances in an entirely different manner from how we perceive and measure time intervals. But stay with me as we pursue this idea further.

What would we expect to find in this spacetime? Obviously they would have to be entities that have both a spatial and a temporal aspect, i.e. they will be characterised by both a position in space and a point in time. We are talking about localised **events**. The start of the spacecraft's journey was just such an event: it had a position in space (the position of the Earth) and it occurred at a certain instant of time (the time of the launch). Likewise, the end of the journey was an event, one having a different spatial position (the position of the planet) and occurring at a different instant of time (the time of the arrival). Now, as we have seen, the astronaut and the controller disagree about both the spatial separation of these two events and also their temporal separation. However—and this is the important point—they *do* agree about the distance or interval between the two events in four-dimensional space-time ('distance' or 'interval', call it what you will—physicists use both terms interchangeably). Moreover, *all* observers, regardless of their relative motion, agree about the distance between *all* pairs of events in spacetime.

This universal agreement encourages the view that what is *real* is not spatial distances and time intervals, but this distance in spacetime. Spatial distances are but three-dimensional projections of this four-dimensional reality, and time intervals are but one-dimensional projections. As such, they are nothing more than appearances that will change according to one's viewpoint. In the case of the pencil being held up, a different 'viewpoint' meant sitting in a different position relative to the pencil. In the present context, a different viewpoint—one that involves both space and time—entails a different speed relative to what is being observed (speed being distance/time).

The proposal for a four-dimensional spacetime was originally put forward by Hermann Minkowski, one-time tutor to Einstein. This was two years after the publication of Einstein's theory. Einstein himself later concurred with this interpretation of his findings regarding space and time, stating, 'Henceforth we deal in a four-dimensional reality, not a three-dimensional reality evolving in time.'

FIG 8 An attempt to illustrate four-dimensional spacetime with three fingers representing the spatial dimensions, and the thumb the time dimension.

So that neatly solves the problem of those differing perceptions of distances and times.

Except that the solution we have come up with is one that many do not find very satisfying. Certainly, if your conception of a satisfying explanation is one where you can form a mental image of what is going on, then you are going to be disappointed. As mentioned at the very start of this book, our brain has its limitations. And one thing it cannot do is form a mental picture of four axes all mutually at right angles to each other. I can spread out three fingers and the thumb of a hand and say, 'The three fingers represent the three spatial dimensions, and the thumb represents the time axis' (Figure 8), but it is a cheat: the angles are wrong. No, we have to resign ourselves to the lack of an accurate visual representation of what is going on. But to physicists that does not matter. They can fall back on mathematics. In fact it is quite easy. The normal mathematical expression for a spatial distance consists of three terms—one for each of the contributions from the three different spatial components (up–down, left–right, backwards–forwards). On extending this to four dimensions, all we have to do is to include a fourth term for the contribution due to the time component. That for a physicist is sufficient. But if you are not 'into maths', I can well understand the difficulty you are encountering in getting an initial purchase on this four-dimensional idea. All I can say is: 'Give it time; familiarity helps.'

But having said that, it has to be admitted that there is something about four-dimensional spacetime that causes difficulty for everyone—including physicists. In fact, it is something that, to this day, divides the physics community. We can see the problem in the following way:

This four-dimensional spacetime is sometimes called the **block universe**. It encompasses all of space and all of time. As far as the time axis is concerned, that includes not just the present, but also the past and the future. Spread out three fingers and the adjoining thumb of one of your hands, as I described earlier. As long as you are not too fussy about the angles between them being wrong, this configuration of fingers and thumb represents spacetime, with the thumb, as before, representing the time axis. Choose one point on the thumb—perhaps the knuckle joint. That represents the present instant. A point closer to the base of your thumb represents the time you started reading this book, and another point closer to the thumb nail represent the moment when you finish reading (and take a headache pill). It is all there—each moment of time on an equal footing. According to this way of looking at things, the past and future exist just as much as the present instant!

And this, of course, is completely contrary to our normal way of viewing time, where only the present instant—'now'—exists; the past has ceased to exist and the future has yet to exist. As far as living our lives is concerned, the future is open; it is not fixed. It remains uncertain until we make up our minds what future action to take. But according to the spacetime, or block universe, idea the future is fixed; it is simply waiting for us to come across it. By 'come across it', I mean consciously experience it.

Now let me hasten to say that, although I myself accept the block universe idea, and I suppose the majority of physicists do, there is a significant number of other physicists—including well-known and highly respected scientists—who strongly hold that the block universe does not represent reality. Fair enough, they argue, if one adds a fourth term involving time to a certain mathematical expression (what we have been calling the distance between two events), then all observers will agree about its value, but such mathematical agreement does not necessarily have to correspond to a real, existent four-dimensional entity. It could be just a piece of mathematics.

For all I know, you might be inclined to go along with them; it is a natural sort of reaction. One of the fears of accepting the block universe as a physical reality is that it appears to compromise our sense of free will—the ability to help shape the future. We have already seen how free will seems hard to reconcile with the deterministic workings of the laws of nature. Now it appears to be under attack for a second time. If the future already exists in some sense (and please do not press me to answer 'In what sense?'!) then this might count as another nail in the coffin.

I do not see it like that. What we have in the block universe is a record of our life histories from conception to death—but that is all it is: a record. An analogy would be to think of a video record of your most recent holiday. It is a series of snapshots taken at successive times. At one end of the tape is the scene of you packing the car, then comes the motorway journey, the arrival at the hotel, etc., culminating in the return home. It is all there. Given a copy of your tape, I can dip into it and sample different episodes at will. After several viewings, I know the whole sequence off by heart. Show me any particular frame, and I know what comes next. I know what you and your family will be doing next. It is fixed. However, given the fact that the record is fixed, and that from any one incident I am able to predict what the next will be, does that mean I am now viewing the actions of fully prescribed robots? Of course not. What I am doing is viewing a record of people making what we regard as genuine, free-will decisions that shape what succeeding frames in the sequence will contain. The video record is a record of free-will decisions.

Obviously, the 'you' to be found at any point in the sequence does not have the information of what is to come later in the sequence, only the information about the earlier parts of the sequence. Those earlier events are known, and hence recognised to be fixed, but not those occurring later in the sequence. The latter are experienced as uncertain because, unlike a viewer of the tape, you did not at that time have access to the whole record.

And that is how I believe it is with us in relation to spacetime. At any point in time—like the present moment—we have access to the information contained in the record of our past life, but not to any information contained in the record of our future lives. The future, nevertheless, is all there, etched into the fabric of the block universe. It is a record of all the actions that will arise from decisions freely taken—actions that arise not as a result of us blindly conforming to some imposed preordained scheme, but as true and faithful expressions of the kind of person we genuinely are.

Does four-dimensional spacetime imply that the future is, in some sense, fixed? ≺

Does this in turn compromise our sense of free will? ≺

But that is not to say that I am 100% happy with the block universe. Acceptance of it does seem counterintuitive. I suppose that I accept it primarily because I do not see a viable alternative. 'What's the problem?' you might ask. 'What's wrong with sticking

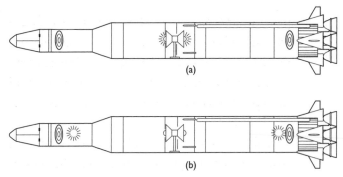

(a)

(b)

FIG 9 An experiment with light beams, as observed by the astronaut.

to the conventional idea that all that exists is the present, the past having ceased to exist, and the future has yet to exist?' In fact, there is a lot wrong with it, and to see why, I must introduce you to another strange consequence of the special theory of relativity.

It arises once again because all observers in uniform relative motion assign the same value for the speed of light. I want you to imagine a spacecraft with a source of light placed dead centre between its front and back walls. The lamp is switched on and simultaneously emits two pulses of light, one directed towards the front of the craft, the other to the rear. What happens to the light?

As far as the astronaut is concerned both pulses travel at speed, c. Both have the same distance to travel, so they arrive at their respective walls simultaneously (Figure 9), which is all very straightforward.

But now consider the situation from the point of view of the controller on the ground watching these events

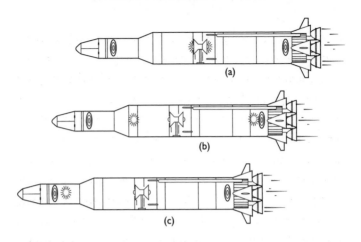

FIG 10 The same experiment with light beams, as observed by the mission controller.

happening in the craft while it speeds past him. When the lamp is switched on, he sees both pulses travelling at speed, c, relative to him—just as the astronaut saw them travelling at speed, c, relative to her (Figure 10a). But then comes a difference. Whereas the astronaut regarded both pulses as having to travel equal distances before reaching their respective walls, this is not the case for the controller. In the time it takes for the rear-going pulse to reach its wall, the wall moves forward to meet it. This reduces the distance to be travelled and hence reduces the time of passage (Figure 10b). Meanwhile the pulse moving forward has to chase after the retreating front wall. It therefore has further to travel and takes a longer time to catch up (Figure 10c). Thus, according to the controller, the rear-going pulse reaches the rear wall before the front-going pulse reaches the front wall. In other words, the arrival times of the pulses are *not* simultaneous. This phenomenon is known as the **loss of simultaneity for events separated by a distance**. A rather long-winded expression I admit, but it is essential to add the phrase 'separated by a distance'. This is because both observers are agreed on the simultaneity of events occurring at the same place, for example, the emission of the front-going pulse being simultaneous with the emission of the rear-going pulse from the lamp at the middle of the craft (Figures 9a and 10a). It is only when dealing with events separated by a distance—in this case, what is happening at the front and rear walls—that there can be disagreement.

What does this talk of loss of simultaneity have to do with our discussion of the block universe? Simply this: When someone discounts the idea that the whole block universe really exists, and insists instead that it is only what is happening right now that exists, presumably they mean to include what is happening right now in locations other than where they themselves happen to be. If one happens to be in London, say, then I take it one would hold that what exists is not just what is occurring at present in London, but also whatever is occurring simultaneously in New York and elsewhere. But there's the rub: that word 'simultaneously'. For an astronaut flying at high speed over London at that instant, what is happening in New York right now is not what is happening according to an observer on the ground. She would pair off a different set of events in New York to be simultaneous with what is occurring in London. These would either lie in the past or the future of the New York events chosen by the observer on the ground. (Which one it is—past or future—will depend on whether the astronaut is flying towards or away from New York.) Thus, according to the observer on the ground, the New York events chosen by the astronaut to be simultaneous with the present have either ceased to exist, or have yet to exist. In this way we see that to maintain that all that exists is what is happening in the present gets one into deep trouble. Different observers, having different perceptions as to what is occurring at this instant, arrive at contradictory conclusions as to what exists at this instant.

None of this, however, is a problem for adherents to the block universe. What exists in New York? Everything! The differing views as to which events separated by a distance are to be considered simultaneous becomes purely a matter of which events to label as having the same time value. Such a labelling procedure has no physical significance. Observers moving relative to each other have differing procedures for labelling.

On the other hand, if it is a question of *existence*, that *would* be a matter of physical significance. And, as we have seen, the conventional view of time, in which a fundamentally different status is assigned to past, present, and future events, leads to inconsistency and paradox. For this reason, I count myself as a block universe man, albeit a somewhat reluctant one because I really do appreciate other physicists' concerns about it. The physical reality of the block universe appears to me to be another question that could possibly defy ultimate resolution.

> *Does the loss of simultaneity for events separated by a distance invalidate the notion that only the present exists?*

9

The nature of time

What more can be said about time in addition to its having a close connection with space?

Nothing about spacetime changes. It is quite static. For something to change, it must change in time. But spacetime is not in time—on the contrary, time is in it. The notion of things changing—the *flow of time*—stems from our conscious awareness of spacetime. Consciousness is sometimes described as directing a searchlight beam onto the time axis of the block universe. The point it picks out is what we call the instant 'now'. We are consciously aware only of this one instant, and that it is why it is singled out to be special. As far as physics is concerned, however, there is nothing at all special about it; it has no different a status from any other instant of time. No sooner has the searchlight beam of consciousness identified this particular instant of physical time to be labelled 'now' than it moves on, in the direction labelled 'future' to alight on the next instant of time. And so on.

We can think of there being two types of time. The physical time belonging to spacetime, which we measure with a clock, and the mental time of consciousness. The latter has its own ways of being measured—or at least of being roughly estimated. If we sit still, with our eyes shut, so we cannot see a timepiece, we are still aware of the passage of time. We become aware of how long we have been sitting there with our eyes shut—at least to the extent of knowing whether we have been doing it for a short time or a longer time. We might even hazard an estimate as to how long we have been doing it in terms of seconds and minutes. We can then, on opening our eyes, compare that internal mental estimate with how much time has passed in the physical world according to a reading on a clock. Sometimes we underestimate the physical time, whereupon we talk of how 'time flies'; on other occasions (particularly if we are bored) time appears 'to drag'.

How we make internal mental estimates of time is not understood. It might have something to do with the progressive fading of distant memories. So if our recollection of an event is somewhat hazy it is likely to indicate that a significant passage of time has elapsed since then. Another suggestion is that the estimate might be related to how much has happened since the event in question. If, for instance, one knows that one has had many different thoughts, feelings, and emotions since deciding to sit still with one's eyes closed, then that too is an indication of how much time must have passed in order to accommodate so many mental experiences. Perhaps the mind is unconsciously keeping track of various bodily sensations, such as one's heart beat or breathing, and these provide quantitative guidance as to how much time has passed. These are but guesses. We simply do not know how we do it.

But are we right in saying that there are two types of time, one physical, the other mental? Again opinion is divided. I myself think that we do have to deal with two types of time. My reason for saying so is as follows: As already stated, we experience a flow of time. What is 'a flow'? It is a change of something in a given time. The flow of a river is a measure of how much water has passed a chosen point in a given time. But what do we mean by a flow of time? How much time has passed in a given time? That does not make sense—not unless we are talking about two types of time. How much physical time, as measured on a clock, has elapsed in a given span of internally assessed mental time? That *does* seem to make sense.

Mental time is marked by a succession of experiences and thoughts we have had. The sequence stretches backwards a long way before it becomes shrouded in 'the mists of time'. But at the other end, the sequence comes to an abrupt halt. This is the experience or thought that we are having right now. It is only in mental time that the instant *now*, marking the end of the sequence and being the focus of our current attention, that takes on special significance. It is associated with a particular instant of physical time, as we read off a clock but, as previously mentioned, there is nothing especially noteworthy about one particular instant of physical time compared to any other—according to the block universe idea. It is only the searchlight beam of consciousness that fleetingly singles it out for special emphasis as it scans for ever onwards.

If you have found this discussion of two times somewhat confusing and unsatisfactory, then all I can say is 'Good. *I'm* confused; *everyone* is confused. Welcome to the club!' The only people who are happy with the concept of time are those who have not thought deeply enough about it. Of all the questions raised in this book as candidates for being unanswerable, the question 'What is time?' and the associated questions concerning the block universe and the two times, must surely rank as among the most intractable.

> *What is time?*

> *Does the perceived flow of time require there to be two types of time?*

One final aspect of time before we leave this topic: the **arrow of time**. By way of introduction, I want you to imagine that you are viewing a film—a rather bizarre film. It begins with a cloud of dust swooping down upon a pile of rubble lying on the ground. The cloud disappears from sight, and the rubble leaps up into the air spontaneously. It comes together miraculously to form a factory chimney. Flames and smoke are sucked into a hole at its base. The hole seals itself, and we end up with an intact chimney—intact apart from some sticks of dynamite sticking out of its base. I said it was bizarre!

Of course, you will already have guessed what is going on. Play the film the other way round and we begin with a chimney and dynamite sticks. There is an explosion sending out flames and smoke from the base. The chimney collapses to a pile of rubble sending up clouds of dust which gradually disperse. Such behaviour makes sense. Clearly this is the correct way of playing the film. The fact that we can tell which way the film ought to be played, i.e. which is the correct way to show the direction of time, is what we call the arrow of time. With regard to any event we are able to distinguish which other events in the sequence lie in its past, and which in the future of that event.

Or can we? Let us view another film sequence. This time the camera is placed above the centre of a snooker table, pointing directly down. Two snooker balls enter the field of vision travelling at the same speed—one from the top of the frame, the other from the bottom (Figure 11a). They collide in the middle and go out of frame in opposite directions, left and right respectively (Figure 11b). Now the film is shown in reverse. The two balls enter from the sides of the frame (Figure 12a), collide in the centre, and go out in opposite directions, top and bottom (Figure 12b). Which was the

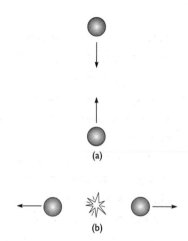

(a)

FIG 11 A filmed sequence of a
collision between snooker balls.

(b)

correct way to show the film? Provided that the balls show
no detectable signs of slowing down, it is impossible to tell.
Both renditions of the collision look perfectly feasible. There
is nothing to choose between them.

What of a third film, this one showing the swing of the
pendulum of a grandfather clock: left, right, left, right...?
Play it backwards and we get right, left, right, left... Again it
is impossible to tell which is the correct way to show the
film; both alternatives look equally plausible.

So what? You might ask. Some processes (a chimney
collapsing) look OK one way but not the other, whereas
others (snooker balls colliding, and a pendulum swinging)
do not. The intriguing point is this: Complex objects such as
chimneys and rubble are made up of atoms and molecules.
What do atoms and molecules do? They collide and bounce
off each other, and they vibrate. If we were able to film two

105

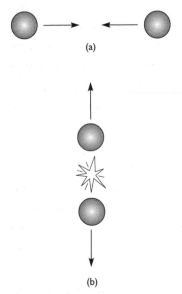

(a)

(b)

FIG 12 The same film sequence shown in reverse.

atoms colliding and bouncing off each other (like the snooker balls) we would not be able to tell which was the correct way to show the film. Physicists call this property of fundamental particles **time reversal invariance**. The same would go for a film of an atom vibrating. Again (as was the case with the vibrating pendulum) there would be nothing to distinguish one direction of time from the other.

So, we have the situation where we put together a whole lot of individual atomic processes—none of which gives us any clue as to the direction of time—and we end up with a chimney sequence where we *do* know the direction of time! Somehow the arrow of time has sneaked its way in. But how?

In fact, there is no mystery. Let us return once more to the snooker table. Earlier we described a pair of balls coming into view from the top and bottom of the frame, colliding in the middle, and exiting left and right (or it might just as well have been the other way round—we couldn't tell). We continue to watch the film. A second pair of balls enters, this time from opposite corners of the frame. They exit left and right. They are followed by further pairs, coming in from random directions, but always leaving left and right— exactly (Figure 13). An odd coincidence? What is going on? It seems as though something is 'sucking' them out along those left—right directions. But, in reality, we know of no such behaviour.

Now we play the sequence in reverse. This time the balls enter from left and right, along the exact same paths, and then leave in random directions. This *does* make sense. We deduce that there are two players positioned out of shot, one to the left and the other to the right. They are deliberately aiming the balls at each other, which then scatter randomly. From this we learn that it is possible to take a whole bunch of events, none of which individually contains any information on the direction of time but, from the way the events *are oriented relative to each other*, we *can* get to know the direction of time. It is not that the reverse sequence is impossible. A relative orientation such that the balls come from random directions and all end up going in the same direction is clearly possible, but unlikely—highly unlikely. It is simply that an orientation of collisions the other way round is more probable.

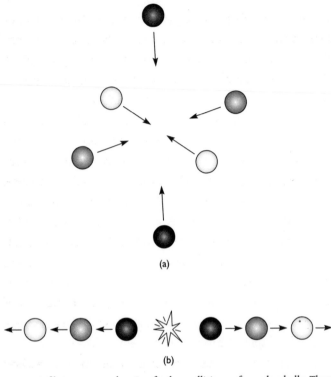

(a)

(b)

FIG 13 A film sequence showing further collisions of snooker balls. They come in from random directions, but always leave exactly left–right.

And this is what is happening at the atomic and molecular level during the chimney sequence. It is far, far more likely that atoms, at the start, in the shape of a chimney plus sticks of dynamite, will end up as rubble and dust, than the other way round. Nevertheless, the reverse is possible. It has to be. Time reversal invariance guarantees that.

But how is the reverse possible? you might ask. How can bricks spontaneously jump into the air? To see how, first consider what happens the right way round: The impact of a fallen brick pushes the surface atoms of the ground vertically downwards. This concerted motion then gets randomised, the atoms jiggling about in all directions and hence warming up the ground a little. The brick's original kinetic energy of motion has now been converted into heat energy. On reversing this process, we would commence with a brick lying on the ground; the ground is warm with the atoms vibrating in random directions. But then, purely by chance, the vibrations all happen to go in the same upwards direction at the same time, this giving the brick an almighty jolt which kicks it flying into the air. The brick eventually lands on top of other bricks similarly propelled upwards. Thus the chimney spontaneously builds itself. Not that it would be wise for any building firm to rely on such a method of construction. The whole process as described is absurdly unlikely, of course. But it *could* happen.

The fact that time has an arrow is where the **second law of thermodynamics** comes in. There are various ways of expressing this law, but essentially it is saying that disorder increases with time. A chimney is an ordered state of matter; a higgledy-piggledy pile of rubble is not. Because of the degree of order, there are relatively few orientations of the bricks that would constitute a recognisable chimney, but there is a vast—indeed limitless—number of orientations of the bricks that would make up a disordered pile of rubble. Suppose we are shown two photographs, one showing a chimney, and the other a pile of rubble, and are asked which came first. We know that one of the relatively rare combinations of atoms known as 'a chimney' is much more likely to end up as one of the limitless number of combinations known as 'a pile of rubble', than the other way round. If one started out with one of the combinations known as 'a pile of rubble', that would almost certainly be succeeded by one of the other many combinations known as 'a pile of rubble' rather than by one of the rare combinations known as 'a chimney'.

But you might argue that disordered piles of bricks do sometimes become chimneys. True, but this tends to happen only in the case of **open systems**—meaning only for situations that are open to influence from outside. In the present case, it is the outside influence of construction workers. The second law of thermodynamics applies only to **closed systems** where one takes into account all relevant factors. The bricks and cement do not assemble themselves into a chimney. It is done with the help of builders. The builders use up energy. The energy comes from the breakdown of food and the conversion of oxygen into carbon dioxide through breathing. The food was grown with the help of an input of energy from the Sun. The dissipation of heat and light from the Sun is a process causing a great increase in disorder. (This can be readily appreciated if you were to think of the reverse: heat and light scattered randomly throughout space all converging by chance, in an orderly fashion, onto a single location—the Sun.) Thus, to arrive at an effectively closed system—one to which the second law can legitimately be applied—we must take all these factors into account, including what is happening to the Sun. Once we have done that, we can rest assured that *overall* the second law has been obeyed and there has been a net increase in disorder. As long as the overall closed system operates in accordance with the law, localised increases in order, such as the building of a chimney, are permitted.

Now all of this, strange as it sometimes seems, is well understood. I wouldn't want you to think that there is some deep mystery here—one that brings us up against the boundaries of the knowable. Nevertheless, armed with this understanding of the operation of the second law, if we probe a little deeper we do come across something that can give us pause for thought. It takes us back to our consideration of the anthropic principle—the observation that the universe is extraordinarily well suited for the development of life.

Suppose we were to create a universe completely from scratch with no preconceptions whatsoever—no need to think necessarily of starting off with a Big Bang, for example. What would it be like? If the universe were to be one where certain states were more likely than others, then it might seem reasonable for it to start off in one of the more likely states. If the universe were to be made up of separate components (call them 'atoms') then the most likely state would seem to be one of the many states consisting of a more or less uniform continuum of atoms moving about randomly. But had the universe come into existence like that it would have carried on indefinitely looking much the same. How boring!

Fortunately, our universe had no such beginning. Everything started off together, very hot, and expanding. This is a highly ordered state. Only in the far off future does it degenerate into the **Heat Death**—the expected future condition of the universe where everything is spread out, cooled down, and with little of interest happening. Meanwhile all sorts of remarkable developments have taken place on the road from extreme order to disorder—*us* for instance!

What is interesting about the initial state of the universe was the sheer potentiality invested in the Big Bang scenario. Where human creativity is concerned, one aims at producing a completed product—be it a painting, sculpture, symphony, novel, poem, etc. The basic outline idea for the final product is there from the beginning. In contrast, what marks the creativity of the universe is an in-built potentiality for realising objects of interest—a potentiality that comes to light only over the passage of time as it evolves away from its initial unusual state to one of the more commonplace states. The world could have been so different. A lucky happenstance, perhaps? If so, perhaps it should be included in the list of 'coincidences' that make up the anthropic principle.

How are we to understand the built-in creativity of the physical world? ≺

We have already noted that there was nothing amiss about the pile of bricks and bags of cement being converted into the ordered form of a chimney.

It arose out of the conscious efforts of the builders working to a predetermined set of plans. How about the human body? It too is derived from basic raw materials—food, water, and oxygen. Once again order is being created on a localised scale—but fully in accord with the second law of thermodynamics when all factors are taken into account. But this time where is the predetermined plan? Where are the builders consciously working towards a preconceived end?

The theory of evolution by natural selection shows that the human body has evolved over time from more primitive ancestors, and they in turn had evolved initially from inanimate chemicals—the so-called primordial soup. It all happened of its own accord. It required no conscious thought or goal-oriented planning. It was the result of natural causes. It came about as a result of what is sometimes called the **self-ordering nature of matter**. This is the ability of the constituents of matter to come together and, of their own accord, produce structures having a wide variety of shapes and properties. Take, for example, the simple case of hydrogen and oxygen. Together they form a highly flammable mixture, but two atoms of hydrogen, when combined with one atom of oxygen produce water—a liquid used for putting out fires. Or take the case of common salt—sodium chloride. It results from combining an explosive metal to a poisonous gas.

Things could have been so different. It doesn't take much effort to imagine a world where the basic constituents of matter might have been simple spheres—hard spheres like snooker balls. Sounds reasonable enough, but what would such a world be like? Again very boring. The same if the constituents had been hard cubes. Even allowing for the spheres, or cubes, to be sticky, and thus capable of bonding together to form larger structures, such structures would be nothing more than shapeless lumps.

But that is not the way it is with our world. Without the need for any intelligent agency to take control, the natural outworkings of the laws of nature, over the 13.7 billion years that have elapsed since the Big Bang, have out of their own subtle nature, produced the double helix structure of the DNA molecule and all the other features that go to make up the bodies of humans and the other animals. Is it not remarkable that in a world governed by the second law of thermodynamics, there can be pockets of activity, defiantly swimming against the inexorable tide sweeping all things towards disorder, in which something as complicated and ordered as the human brain can emerge?

Not, let me repeat, that this in any way compromises the second law, once one takes into account all relevant factors— what is happening in the Sun, for instance. Indeed, scientists fully understand the mechanics of how this self-ordering property of matter works. With the help of quantum theory, one can understand the structure of atoms, why there are 92 different naturally occurring types of atom, or element, how they bond together, and how in doing so they give rise to molecules that can have very different properties from those of their constituent parts. Crudely speaking, atoms are more like Lego blocks than spheres or cubes—92 different types of Lego block that have the potential for fitting together to form all kinds of interesting shapes. Without that self-ordering characteristic of the atoms we would not be here. For that reason perhaps this is yet another feature we ought to add to the list that constitutes the mystery of the anthropic principle.

> *What significance, if any, ought we to attach to the self-ordering nature of matter?*

10

High energy physics

We have already said something about the structure of atoms, but now it is time to probe deeper into the nature of matter.

As is commonly known, atoms consist of a central **nucleus** surrounded by smaller particles—**electrons**. The nucleus carries positive electric charge, while the electrons have negative charge. The like charges on the electrons give rise to a mutually repulsive force while the unlike charges on the nucleus and the electrons leads to an attraction that holds the electrons in the vicinity of the nucleus. But not too close; the atom is mostly empty space. If we envisage the nucleus modelled by a golf ball sitting on a runway at the centre of London Heathrow airport, the electrons would be found to roam out as far as the perimeter of the airport.

The atoms of the different elements vary in the number of electrons they have. Hydrogen has one, helium two, lithium three, etc. all the way up to uranium with 92.

The electron is believed to be fundamental, i.e., it has no internal structure; it is not made up of component parts. Not so the nucleus. Two nuclei colliding with each other can lead to their break-up. This splitting of the nucleus reveals that it is made up of constituent particles which we call **nucleons**. There are two types of nucleon: the **proton** and the **neutron**. They are very similar to each other, except that the proton has equal but opposite electric charge to the electron, while the neutron has no charge. It is the presence of the protons that gives the nucleus its positive charge. Thus a second distinguishing characteristic of an atom (assuming that it is in its normal state of having zero charge overall) is that it has the same number of protons in its nucleus as there are electrons outside the nucleus.

When two or more atoms get near each other, their electrons can feel the attraction of the other atom's nucleus as well as that of their own, and this can result in two or more atoms sticking together and thus forming a **molecule**. There are hundreds of thousands of different known molecules (or chemicals). The familiar force we experience when pushing everyday objects is another manifestation of the same electric forces. On throwing a ball, for instance, one has to think of the outermost electrons of the atoms making up the hand as electrically repelling the outermost electrons of the atoms of the ball.

We have seen how the positive charge on the proton holds the negatively charged electrons together in the atom. But you would expect the like charges of the protons to lead to a repulsive force between them—a force tending to blow the nucleus apart. The fact that they are bound together tightly calls for a second type of force—a force of attraction so strong that it is able to overcome the mutual electric repulsion between the proton constituents, leading to an overall attraction. This force is called the **strong nuclear force**. In contrast to the electric force, which acts over large distances (falling off as the inverse square of the distance), this additional force is short-ranged; it is experienced only by those protons and neutrons that are close neighbours. As one imagines building up ever bigger nuclei by adding more and more protons and neutrons, the strong attractive force experienced by a proton situated on the surface of the nucleus hardly changes. This is because, as far as the short-ranged strong force is concerned, it feels only the influence of its near neighbours—those within the force's range. It is unaffected by protons being added elsewhere. The long-range repulsive electric force, on the other hand, goes up with each new proton added. Thus there comes a point where the repulsive electric force proves too much for the strong attraction and any further protons added do not stick. This limit comes when the number of protons present reaches 92. Hence the number of naturally occurring elements.

So are neutrons and protons fundamental, or are they, in their turn, made up of yet more basic constituents? How could we find out? The obvious answer seems to be to open them up. After all, we broke open molecules to find the atoms; the atoms were broken apart to reveal the nucleus and the electrons; and the nucleus was broken up to show the neutrons and protons. And so it is that one moves into the realm of high energy nuclear physics where we accelerate charged particles—either electrons or protons—and slam them into targets, and watch what happens when the projectiles collide with the nucleons belonging to the target.

At least that is what we do today. Way back in the 1950s we did not have high energy accelerators. In those days we had to make use of high energy particles coming from outer space—**cosmic rays**. The trouble was that most of them collide with nuclei high up in the Earth's atmosphere; most do not get down to sea level. So it was that we had to lug our detecting equipment up mountains or fly them in balloons. Being based in England, which does not have a mountain worth speaking of, the team I belonged to had to look abroad. We set up our research station on Mount Marmolada in the Dolomites of Northern Italy.

And what did these cosmic ray experiments show? Did we manage to observe a proton smashed open? No. The protons remained intact regardless of the violence of the impact. Instead we created other new particles—particles that did not exist before the instant of the collision. As previously noted, it was not a case of getting something for nothing. Although new matter is created, some of the kinetic energy originally carried by the projectile particle is now missing. This is a possibility allowed by Einstein's formula, $E = mc^2$, which we encountered earlier.

This process whereby energy is transformed into matter is reversible. In the hot interior of the Sun, as we have already seen, collisions between the nuclei of elements such as hydrogen and helium can give rise to rearrangements of the neutrons and protons leading to the production of new elements with the release of energy. This energy arises through the new elements having less mass than the sum of the masses of the original particles; in other words, some of the matter is lost. A proportion of the energy originally locked up in the form of matter has been converted into heat and radiation—the process known as **nuclear fusion**.

But let us get back to the production of new matter in high energy collisions. We begin by noting that these new forms of matter can be produced only in certain fixed amounts. For example, we can produce a particle called a **pion**. It has a mass of about $273\,m_e$ (where m_e is the mass of the electron). But we never produce a particle of arbitrary mass, say, $272\,m_e$, or $274\,m_e$. It seems that it has to be $273\,m_e$. The pion was found to be but one of the types of particle that could be produced. For a time, it was assumed that a crucial question was to discover how many of these new types of particle there were. As more powerful accelerators were built, thereby increasing the energy available in the collisions, it became possible to create ever heavier new particles. We now know of more than 200. Indeed, there appears to be no limit to the number of different types of particle one can make; it simply depends on how much energy is available for providing their mass.

The new particles can carry electric charge; a positive charge like the proton, or a negative charge like the electron. So does that mean we have created electric charge out of nothing? In a sense yes. The trick is to make sure the *net* charge remains unchanged. A positively charged particle can be produced only if, at the same time, a negatively charged particle is also produced. We talk about electric charge being conserved—meaning net charge.

One of the early puzzling features of the collisions giving rise to these new particles was what came to be known as 'associated production'. Whereas pions could be produced singly, other particles could not. Even though there might be plenty of energy in the collision for providing their mass—and they did not have to be electrically charged so conservation of charge was not a problem—it was found that they could only be produced in association with some other particle. This led to the idea that these new particles might be carrying some hitherto unknown property of matter which, like electric charge, also had to be conserved in the collisions. One could only produce a particle carrying a positive unit of the new property if at the same time one produced a particle carrying a negative unit of this same property. That way the net amount of the property remained unchanged. This new property was given the colourful name **strangeness**—for no other reason than that what was going on struck us at the time as being very strange. It was found that a certain particle was only ever produced in association with two other particles. This behaviour could be explained by assigning -2 units of strangeness to it, thus necessitating the production of two other particles each carrying $+1$ unit of strangeness. Ordinary matter—neutrons and protons—together with the pions that we mentioned, have no strangeness and hence are not subject to the same restriction.

And strangeness is not the only new property of matter that has been discovered. There are situations where a new particle cannot be produced on its own even though there is enough energy to produce its mass, and such a production would violate neither the law of conservation of electric charge, nor that of conservation of strangeness. This points to the existence of yet another new property of matter, one that has also to be conserved in the collisions. I myself enjoyed the privilege of being a member of the international team of physicists that made the first direct sighting of a particle carrying the property, which was designated—again somewhat whimsically—**charm**. Since that time, two further properties have been discovered. These have been given the names **top** and **bottom**.

There is no doubt that the assumption that these new properties exist has considerable explanatory power. It makes sense of the associated production of particles and other aspects of their behaviour, but at no point do we ever say what these properties are in themselves—what kind of 'stuff' they are. The assumption that these properties exist is a device for explaining *behaviour*—and that is all. It helps us to make sense of our observations as to why only certain combinations of particles can be produced.

Now, I can well imagine how you might be feeling at this point. How dare physicists go around *inventing* new properties just to get themselves out of the awkward situation of having to account for associated production. What actually are these properties supposed to *be*? What are they *in themselves*?

We cannot say. When it comes to specifying in more detail what kind of actual stuff these properties are we fall silent. So, how embarrassing is that? No more so than our embarrassment at not being able to explain more familiar properties of matter. Take electric charge. We are all familiar with electric charge. It is what makes clothes cling when taken out of a tumble drier. This electric force obeys an inverse square law such that if the charges are placed twice as far apart from each other, the force becomes a quarter of what it was; 10 times further away, 100 times weaker. Electric currents flowing through wires power all the electrical gadgets in the home. An electric current passing through a wire is explained in terms of electrons passing along the wire. Such moving charges produce a magnetic field around the wire. This can be checked by placing a pocket compass close to the current-carrying wire and watching the needle move. The Earth's magnetic field is produced by the flow of electrically charged particles in its core. One of the greatest triumphs of physics is our understanding of electromagnetic phenomena such as these. We have Clerk Maxwell to thank for this. As we all know, modern-day life has become dependent on the applications of that knowledge in so many ways— as is readily demonstrated during the black-out caused by a power failure. The sophistication of today's electrical gadgets, computers, etc. testifies to how well we understand electricity and can manipulate it to our own ends.

But that said, what actually *is* electric charge? All we have talked about is what it *does*—the *effects* produced by electric charges. If we assume that objects have this thing called 'electric charge' then this is how they are going to behave. But we never say what the thing *is*—not what it is in itself.

This inability to go further and account for what something is in itself is not confined to electric charge—by no means. It applies to *all* properties of matter. Take, for example, *mass*. As we learn from *Newton's second law*:

$$\text{force} = \text{mass} \times \text{acceleration}$$

If we pull on different objects with the same force (for instance, the force exerted by a spring stretched by a given amount) we note that the objects accelerate, but to different degrees. Why the difference? We say that the objects have mass. It is the amount of mass possessed by the object which determines how it responds to the force. Half the mass receives twice the acceleration; a third of the mass, three times the acceleration. All very simple and straightforward. But note that, once again, all we are talking about is behaviour—how objects are observed to move in space and time. At no point do we ever specify what kind of stuff the 'mass' property is supposed to be.

Not that there is anything new in this. The philosopher Immanuel Kant in his book *Critique of Pure Reason*, first published in 1781, claimed that things-in-themselves are not cognitively accessible, and no verifiable description of their nature can be given. All we can deal with are the observational phenomena to which they give rise. Such thinking was foreshadowed by early theologians in considering what could and could not be said about God. The Jewish philosopher, Philo (c.20BC–AD45) drew a distinction between the 'essence' of God, which is absolutely unknowable, and his activities in the world, his 'energies', which were all that were accessible to our understanding. In the fourth century AD, St Augustine said much the same when he claimed that we could not know 'the name of the substance of God'.

But back once more to physics. It was this same inability to talk about things-in-themselves that we encountered when trying to explain the true nature of space and of time. With regard to space, for example, all we could say is that objects were observed to behave in space as if space were some mysterious something that could be curved, as if it was a frothing sea of unseen virtual particle pairs popping into and out of existence, as if it consisted of a sea of negative energy electrons, etc. As for the nature of time...!

When it comes to understanding things-in-themselves—whether it be space, time, or properties of matter—we do indeed appear to be up against a barrier of the knowable. We do not have a language that has terms that would be capable of describing the nature of the 'stuff' of a thing-in-itself. And this lack of a suitable language stems from the inability of the human mind to conceive of such things. We just have to learn to live with this limitation. And that is what we are doing when physicists propose new properties such as strangeness, charm, etc. Yes, at a deeply fundamental level, they elude our full comprehension, but such proposals do have explanatory value at the practical level of understanding outward behaviour patterns, and for that reason they are useful.

The problem of ≺
understanding
things-in-
themselves.

Having seen how we can produce particles with these new and unfamiliar properties, it might have occurred to you to wonder what subsequently happens to these exotic forms of matter. Are we progressively accumulating piles of them? No. All the new particles are unstable. Straight after they are produced, in a mere fraction of a second, they decay into ordinary matter: neutrons, protons, electrons, and neutrinos. How do they do this?

It is at this point that we need to say something about the ways in which these different particles interact with each other. There are various types of interaction or force. (The terms 'interaction' and 'force' are used interchangeably to describe any way in which the particles exert an influence on each other. This applies not only to changes in motion—the day-to-day context in which we think of forces operating—but also to such interactions as lead to a change of identity of particles.)

First one ought to mention the gravitational interaction which operates between all particles. This is the weakest of all the forms of interaction. That sounds an odd thing to say because we are all familiar with the idea of powerful gravity holding us to the surface of the Earth, the Earth in orbit about the Sun, and especially the manner in which it can exhibit its strength in black holes. But in fact the electromagnetic force is much more powerful. The reason why this is not obvious is that electric charge comes in two forms: positive and negative, and this gives rise not only to attraction but also to repulsion. The electric forces holding the atom together are extremely powerful—far more so than the gravitational attraction between the electrons and the nucleus. But the electrical force exerted on something placed at a distance from the atom is hardly noticeable because the forces exerted by the charges on the nucleus and electrons act in opposite directions and largely cancel each other out. With the gravitational interaction, however, there is no cancellation; it is cumulative. Hence it is gravity that holds the Earth in orbit about the Sun, and not an electrical force. In the context of what is happening in high energy collisions at the subatomic level, we can ignore gravity.

So we must take note of the electrical force. But this alone cannot be responsible for what happens in high energy collisions. After all, some of the particles taking part in these interactions are uncharged and therefore unaffected by electric forces. Thus we need to call upon another type of interaction—the strong nuclear force that we came across earlier in connection with our discussion of how the nucleus is held together. Because of the strength of this form of interaction it acts rapidly when another particle comes within its range. A proton projectile from an accelerator travels close to the speed of light (3×10^8 metres per second). If it is to produce new particles, it has to interact in the time that it takes for it to cross from one side of the target proton to the other—a distance of 10^{-15} metres. That takes of the order 10^{-23} seconds. So that sets the scale for the strong interaction.

So, in high energy physics we must take note of both the strong and electromagnetic interactions. But there is a third type of interaction: that responsible for the decay of the newly created particles formed in the collisions. It is an interaction responsible for the radioactive decay of nuclei arising from the transformation of a neutron into a proton, an electron, and a neutrino. We call it the **weak interaction**. It is designated 'weak' because it takes a comparatively long time to happen. Whereas, as we have seen, the strong interaction occurs in typically 10^{-23} seconds, the weak interaction takes much longer; for example, a charged pion takes something like 10^{-8} seconds to decay.

That is one difference between the strong and the weak interaction. Another is that, although both obey the laws of conservation of electric charge, the weak interaction is no respecter of certain other conservation laws such as those of strangeness, charm, top, and bottom—the new properties that were discovered in high energy collisions. That is how the newly formed particles carrying these properties revert back—through weak interaction decay—to ordinary matter like protons and neutrons.

So, there we have it: over 200 new particles to deal with, and new properties to add to the more familiar ones, and different types of force operating between them. It is all getting rather messy and complicated! You might be tempted to think that, with these new types of particle and their odd properties not occurring naturally, and when produced in the somewhat artificial conditions of high energy collisions they almost immediately disappear again, can we not quietly forget about them as being of no particular significance or lasting interest? Why not just stick with our good old neutrons and protons? No, we cannot do that. It turns out that these new particles, the existence of which is so fleeting, are close cousins of our more familiar types of matter. Just as one might learn something about an individual person by seeing them in the context of their immediate family, so a study of the neutron and proton in the context of these new particles proves to be the key to understanding their true nature—their elusive inner structure.

This comes about in the following way: In 1962, Murray Gell-Mann and Yuval Ne'eman, independently of each other, recognised that the particles could be classified in a revealing manner. By choosing particles that had certain properties in common and plotting them out according to two other properties, various symmetrical patterns, known as **SU3 patterns**, emerged. Figure 14 shows the forms of two of

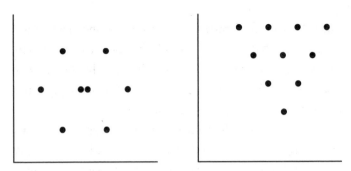

FIG 14 Typical SU3 octet and decuplet patterns.

them: a hexagonal grouping of eight particles with two at the centre, and a triangular one of ten. The '3' in the name SU3 is related to the three-fold symmetry obtained by rotating these patterns through 120°. (And, in case you are wondering, the SU stands for *special unitary*.)

The situation facing Gell-Mann and Ne'eman bore similarities to that which had earlier confronted Mendeleev when compiling his periodic table of the atomic elements. Recall how Mendeleev had displayed systematically recurring patterns of behaviour by dividing the elements up into groups. At first not all the groups were complete; there were gaps. He rightly assumed that these pointed to the existence of hitherto undiscovered elements—elements with properties that were predictable from where they should appear in the pattern. Likewise, Gell-Mann and Ne'eman found a gap in one of their patterns—the bottom place of a triangular decuplet like the one shown in Figure 14. It corresponded to a hypothetical particle that would be negatively charged (there would be no positive or neutral version) and would carry an unprecedented -3 units of strangeness. They boldly predicted that there would be such a particle, and named it the Ω^- (omega-minus). Subsequently, the Ω^- was discovered with just those properties. This was a great triumph for the SU3 mode of classification.

Mendeleev's periodic table had a deeper significance than just being a handy method of classification. In addition, it hinted at the inner composition of the elements—an inner structure that was responsible for the observed relationships. It suggested that the elements were to be regarded as variations on a common theme: a central nucleus surrounded by electrons.

In 1964, Gell-Mann and George Zweig likewise suggested that the similarities and family patterns displayed by the new particles were a reflection of some inner structure. This proposal held out the possibility that the 200 or more particles, until that time regarded as 'elementary', were in fact composites constructed from yet deeper fundamental constituents. These constituents were to be called **quarks**. They are treated as being point-like, having no inner structure consisting of 'subquark' constituents.

Are we *sure* that there is not another Russian doll inside? For a time, it was suggested that quarks in their turn might be made up of even more basic particles, provisionally given the name **preons**. A few physicists worked on this idea in the late 1970s, but this is now no longer an active field of enquiry. It is believed that quarks *are* truly fundamental. However, the question of whether the quarks really are point-like is very much a live issue. We shall be saying more of that in the final chapter when we come to consider string theory.

It is found that there are six types of quark, each having its own characteristic properties. One of them, for example, carries one unit of strangeness. So if this type of quark is a constituent of one of the particles we have been considering, then that particle will also exhibit the property of strangeness. Similarly, another type of quark carries a unit of charm, and is a constituent of the charm particles, and so on. The same holds for the other new properties, top and bottom. The fact that neither the neutron nor the proton possesses any of these new properties implies that they do not contain those kinds of quark. In addition to the six quarks, there are six **antiquarks** possessing the opposite values of the properties.

The form of the SU3 patterns indicates that the particles emerging from the collisions consist either of three quarks (q,q,q) bound together, or as a quark/antiquark (q,$\bar{q}$) pair. The neutron and proton, together with the aforementioned Ω^- are examples of the (q,q,q) combination, while the pion is an example of a (q,$\bar{q}$) pair. The antiproton consists of a ($\bar{q}$,$\bar{q}$,$\bar{q}$) combination (Figure 15).

So far, we have talked of quarks bound up in the particles taking part in the collisions. What of free quarks? Despite strenuous efforts, none has ever been seen. Even in the highest energy collisions, quarks are never ejected. This calls for an explanation.

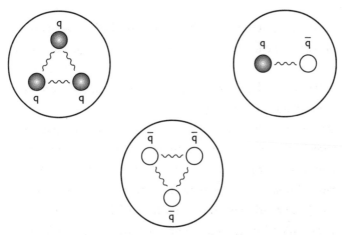

FIG 15 The allowed combinations of quark and antiquark.

One idea canvassed for a time was that quarks were not real: they were mere mathematical entities—useful fictions. Although the SU3 patterns were consistent with the particles being made up of quarks, that was not proof of a quark structure. The particles behaved *as if* they consisted of quarks, but there was no such thing as an actual quark. But then came a conclusive demonstration of their reality. It was an example of history repeating itself. Back in 1911, Lord Rutherford demonstrated the existence of the nucleus by firing projectiles (alpha particles, i.e. helium nuclei) at atoms, and observing that some rebounded at large angles. This indicated that the projectiles had struck a small con-centrated target (the nucleus) within the atom. In 1968, it became possible to fire high energy electrons into the *interior* of the proton. Evidence began accumulating that the elec-trons were occasionally suffering large sideways kicks, indicative of them having rebounded from some small and concentrated electric charge inside the proton. This was confirmation that the quarks were real. Indeed, from the frequency of the large-angle scatters, it could be calculated that there were three quarks inside the proton.

So, having established that the quarks are definitely in there, why do they never come out singly? We begin by recalling how the attraction between the proton and electron of a hydrogen atom arises out of the electromagnetic force operating between the electric charges carried by the proton and electron. By analogy therefore, we are led to introduce an additional kind of 'charge'. We postulate that quarks carry this kind of charge (in addition to electric charge), and that the force binding quarks together arises because of interactions occurring between these charges. For rather complicated reasons we call it **colour charge**, and the interquark force is known as the **colour force**.

It is this colour force that provides insight into the nature of the strong interaction that we encountered earlier between protons and neutrons—the force of attraction opposing the electrostatic repulsion between the positively charged protons in a nucleus, resulting in the nucleus sticking together. Recall how atoms form composite molecules in spite of being themselves electrically neutral. As we saw previously, the electrons of each atom rearrange themselves so that they are partly attracted to the nucleus belonging to the other atom. Thus is generated an external remnant force capable of binding the atoms together. In the same way, the quarks within a nucleon can adjust themselves in such a way as to produce an external remnant force capable of attracting the constituents of the neighbouring nucleon. Thus we see that the strong force operating between nucleons is to be regarded as a manifestation of the powerful and more fundamental colour force operating between the constituent quarks.

An interesting feature of this colour force is that, unlike the electric and gravitational forces, which fall off in strength with increasing separation, the strength of this colour force remains constant as the quark separation increases.

One way of explaining the absence in nature of free, isolated quarks is to say that quarks *can* in fact be knocked out of, say, protons. The trouble is that, in order to do this, one has to hit the quark so hard—with so much energy in order to snap its constant-strength bond to the other two quarks—that one cannot help but produce a quark—antiquark pair at the same time. (The situation is rather similar to that in which one might snap a bar magnet in an attempt to isolate the two magnetic poles—only to discover that one has in the process created a new north and south pole.) The new quark joins the two remaining quarks to reform the proton, while the new antiquark goes off with the knocked-out quark as a $(q,\bar{q})$ combination. In other words, all one has done is to account for the production of a particle such as the pion!

It should be pointed out that not *all* particles are made of quarks. Those that are, we call **hadrons**—a word meaning strong. Hadrons feel the strong nuclear force. (Hence the name of the largest accelerator yet built: the Large Hadron Collider, at CERN, in Geneva.) Other types of particles, such as the electron and the neutrino, are not composed of quarks, and thus do not experience the strong force. They are collectively known as **leptons**—a word meaning light. Thus we have two broad classes of particle—hadrons and leptons—both of which experience the weak interaction, the gravitational interaction, and, if electrically charged, the electromagnetic interaction. But only hadrons additionally take part in the strong interactions arising from the colour force between their constituent quarks.

Summarising so far, we have six types of quark. They are the constituents of the hadrons, and are believed to be truly fundamental. Then there are the leptons. There are six of these too: the electron and two other charged particles, called the **muon** (denoted by μ) and the **tau** (τ). Each of these has its own type of neutrino—thus yielding six leptons in all. Like the quarks, the leptons are regarded as truly fundamental, i.e. they cannot be thought of as composite structures built up from yet more basic entities.

These twelve particles can be divided into three groupings of four. Each grouping consists of two of the quarks plus one of the leptons and its associated neutrino. The groupings are called **generations** (Figure 16).

Earlier we saw how high energy physicists initially thought it was important to establish how many different types of new hadrons could be created in the collisions. This was later recognised not to be an important question at all,

Generations	1	2	3
Quarks	u d	c s	t b
Leptons	e ν_e	μ ν_μ	τ ν_τ

FIG 16 The three generations of fundamental particles, each consisting of two quarks, a charged lepton, and that lepton's neutrino.

there being an unlimited number of ways that the constituent quarks could be put together in different combinations and energy states. We have established what the truly fundamental constituents of matter are, and that there are three of these generations of them. But this in turn raises what seems definitely to be a crucial question: how come there are this number of generations?

In order to account for the matter we see around us, we require only the particles belonging to the first generation: the two types of quark that are the constituents of the neutrons and protons, plus the electron. One might, therefore, be forgiven for thinking that life would have been much simpler had nature just stopped there!

Can we not at least regard the other two generations as being, in a sense, 'less important' on account of the way they give rise to matter that is unstable? By no means. The distinguishing characteristic of the first generation is merely that it happens to contain the two lightest quarks. The quarks belonging to the other generations, like the ones carrying strangeness and charm, are heavier, and heavier quarks have a tendency to decay into lighter ones with the release of the excess energy. Had the strange quark, for instance, happened to be the lightest quark, then all the matter around us would have been made up of strange hadrons. Under those circumstances, we would have regarded the neutron and proton as examples of unstable exotic particles created only in high energy collisions, and quickly decaying to 'ordinary' strange matter. This brings home the fact that there really is nothing special about our familiar neutron and proton; they are essentially on an equal footing with all the new hadrons we have found.

No, we are stuck with three equally important generations. But why three? At present no-one has any idea.

Why are there three generations of particles? ≺

So much for the constituents of matter—both that which makes up what we see around us, and the unstable forms that arise as a result of the high energy collisions. Let us now look more closely at the forces operating between them.

In the spirit of quantum physics where, as we shall be seeing in the next chapter, interactions occur discretely rather than continuously, we regard the mechanism by which a force—*any* force—is transmitted from one particle to another as involving the **exchange of an intermediary third particle**.

Recall from our discussion of the origins of dark energy that so-called empty space could be regarded as a sea of virtual particles popping into and out of existence. This is a consequence of the Heisenberg uncertainty principle whereby a certain amount of energy can be 'borrowed' for a short span of time. These energy fluctuations affect not only how we view space but also how we view particles such as the electron. An individual electron is not simply a single particle sitting there in space doing nothing. All the time it is emitting photons—the packets of light we referred to earlier. Not that we ever see these photons. Having emitted the photon, the electron almost immediately reabsorbs it again (see Figure 17). In other words, the electron is behaving somewhat like a juggler throwing up balls and catching them again. The energy required for producing these ephemeral virtual photons is once again provided by the uncertainty principle.

So much for an isolated electron. What if a second electron passes by? It too is performing the same juggling act. As they near each other, a photon emitted by one electron might be caught by the other; the photon is exchanged (Figure 18). Exchanging photons is how electrons, and any other charged particles, interact with each other—how they affect each other—how they exert the electric force on each other.

Basically we can think of electron 1 emitting the intermediary photon in the direction of electron 2, and in the process suffering a recoil—much as a rifle recoils in the

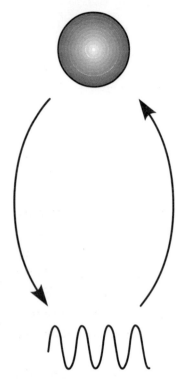

FIG 17 An electron emitting and reabsorbing a virtual photon.

opposite direction to the motion of the bullet. Electron 2, on receiving the intermediary, takes up its momentum, causing it to recoil away from electron 1. The overall effect of this exchange is that both particles are pushed apart. The process then repeats itself when the intermediary photon is returned; there is a further pushing apart. The net effect is that the two particles repel each other; i.e. they experience a repulsive force. This intermediary particle, once exchanged, then disappears, the borrowed energy having been returned.

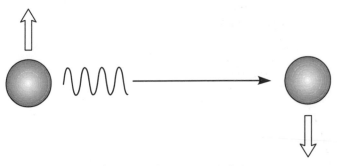

FIG 18 Electrons exchanging a virtual photon.

What about forces of attraction? Essentially the same mechanism, though I suppose this time—if you insist on having an analogy—we must think of the particles throwing boomerangs rather than shooting bullets! Particle 1 emits the intermediary in a direction *away* from particle 2, hence experiencing a recoil *towards* that particle; the latter then receives the intermediary from the opposite direction, and is also pushed towards its companion.

In the case of the electric force between two charges, the intermediary particle is, as we have noted, the photon—a tiny bundle of light energy. The two charges are either repelled or attracted due to the repeated exchange of photons. That being so, it prompts us to ask whether the strong colour force is also open to an explanation in similar terms, the exchange of some kind of intermediary particle between the quarks. The answer is yes, quarks are held together in the hadron by the exchange of particles appropriately called **gluons**. There are in fact eight different types of gluon.

The strong colour force differs from the other types of force we have so far encountered in respect of the range over which it is effective. The gravitational, electric, and magnetic interactions are long-ranged and so give rise to easily observable macroscopic effects—planetary orbits and the emission of radio waves, to mention but two obvious examples. The strong force, on the other hand, is short-ranged, acting over distances of only 10^{-15} metres—those characteristic of the size of the nucleus.

So much for the strong interaction. What about the weak interaction? As mentioned before, it is the weak force that is responsible for the decay of particles. It is also responsible for the interaction of neutrinos. We noted earlier that neutrinos hardly interact with anything; billions of them originating in processes occurring in the Sun, pass through our bodies every second without our noticing them. And yet they do occasionally undergo a collision. They carry neither colour charge nor electric charge, so they cannot interact by either the strong or the electric force. The fact that neutrinos nevertheless do take part in interactions with other particles shows that we must be dealing with another type of interaction— the weak force.

To learn more about it, we must retrace our steps a little. Back in the 1860s, James Clerk Maxwell came to recognise that two types of force, the electric and the magnetic, until then thought to be distinct, were in fact but different manifestations of the same force—the **electromagnetic force**. This process of unifying forces was taken a stage further by

Steven Weinberg (1967) and Abdus Salam (1968), building on earlier work by Sheldon Glashow. They were able to show that the weak force might be a further manifestation of the same force—a combined force that has come to be known as the **electroweak force**. This, at first sight, might appear surprising, the effects of the electromagnetic force being so much more striking than those of the weak force. However, it should be noted that the latter is not weak in the sense of its intrinsic strength being less than that of electric and magnetic forces; it appears weak only in the sense that it operates over an even shorter distance than the strong force: only 10^{-17} metres.

In order for this unification of forces to be possible, the weak force, in common with the other forces that we have been considering, has to be mediated by the exchange of some form of particle. As we saw previously, the basic idea behind this 'exchange mechanism' is that, thanks to the Heisenberg uncertainty principle, one is able to borrow a certain amount of energy for a limited time in order to make the new particle. This particle is then exchanged, whereupon it disappears once more. The lighter the particle, the less energy one needs to borrow, and the longer the time the particle can exist, and hence the further it can travel between the two particles doing the exchanging. In the case of the electromagnetic force, the exchanged particle is a photon. The energy required to produce a photon depends on the wavelength of the light to which the photon belongs; the longer the wavelength of the light, the less energy carried by the photons that make up that light. So by using photons characterised by very long wavelengths, the energy requirement is minimal, so the exchange process can take as long as it likes, and consequently the range of the force is infinite. By contrast, the weak force is short-ranged—this indicating that the exchanged particles are massive and hence they can travel only short distances before having to disappear again. Weinberg and Salam's theory predicted that there would be three of these intermediary particles: the W^+, W^-, and Z^o. At the time of the prediction, no such particles as the W and Z were known. However, the theory was triumphantly vindicated in 1983 by their successful discovery. The W was found to have a mass 85 times the mass of the proton, i.e. $85\,m_p$, and the Z a mass of $97\,m_p$. Why there should be such a large difference in masses between the photon on the one hand, and the W and Z on the other, given that they ultimately come from the same force, is not understood.

The Z^o has proved to be a particularly interesting particle. Like other new particles, it is unstable, its average lifetime dependent in part on how many different decay modes are open to it. The greater the number of different combinations of particles it can give rise to, the quicker it is likely to change into one or other of those combinations. It decays into a neutrino and its antineutrino. So, the greater the number of different types of neutrino, the more decay modes become available to the Z^o, and the shorter will be its lifetime. Its actual lifetime corresponds to there being three different types of neutrino—the three we already know about. From this it follows there are only three lepton doublets (Figure 16).

Not only that but we can, from this observation, go on to make a prediction about the number of different types of quark. We know that the lepton doublets are grouped with the quark doublets to form generations. Thus, if there are only three lepton doublets, there can only be three generations, and hence only three quark doublets. In other words, the number of different kinds of quark is limited to six.

This is important. A disturbing feature of the quarks had been that each newly discovered type was much heavier than its predecessors, ranging in mass from 0.002 m_p to 180 m_p. Heavy quarks mean heavy hadrons to contain them, and the heavier the hadron, the more difficult it is to produce at the particle accelerators because they require more energy. This had caused concern: that there might be quark types that we could never learn about because we physically did not have the resources to build a particle accelerator powerful enough to produce them. However, thanks to the Z^o, this has ceased to be a problem. We now have good grounds for believing there to be only the six already-known types.

This conclusion has a bearing on an earlier question we raised, namely, are quarks truly fundamental, or might they be composites of yet more basic constituents—the 'preons' mentioned before? If quarks were indeed to be constructed out of preons then there might be no limit to the number of different combinations that could be made up, and hence no limit to the number of different flavours of quark. This in turn would mean no limit on the number of different generations, and consequently the number of different types of neutrino. The fact that the decays of the Z^o provide grounds for believing there to be only three types of neutrino appears to rule out the preon idea and leaves quarks as truly fundamental.

We conclude that the complete inventory of elementary particles looks like this:

(i) Six quarks and six leptons;
(ii) Twelve intermediary particles, made up of eight gluons, the photon, W^+, W^-, and Z^o, these being responsible for the forces.

In addition, one should add the antiparticles of the above. Also, we need to mention that there is expected to be a further intermediary particle, the **graviton**, to mediate the gravitational interaction.

This understanding of the structure of matter and of the electroweak and strong forces operating between them is known as the **Standard Model**. It was formulated in the early 1970s. It is an extraordinary achievement of the interplay between, on the one hand, experimental data (the discovery of all the new particles, their properties, roles, and interrelations) and, on the other, the development of the theory. It has so far stood all tests and is universally accepted as our best understanding to date of the nature of matter and how it behaves.

What of the future? We have seen the unification of the electromagnetic and the weak force. It is with a gleam in the eye that one anticipates the possible unification of the electromagnetic and weak force with the strong colour force. This is called the **Grand Unification Theory**, or **GUT** for short. Such a theory holds out the prospect that all three forces should have comparable strengths at high enough energies. At the kind of energies we are accustomed to in the laboratory the forces have different strengths. But the higher the energy, the closer they get to each other. At laboratory type energies, the strong force, as its name implies is very strong. The weak force then comes next (the 'weak' aspect coming from its short range rather than its intrinsic strength), and the electromagnetic force is weaker still. With increasing energy of the collisions, the strong force is found to reduce in strength, while the weak force reduces more slowly, and the electromagnetic force increases. Grand Unification is expected at an energy of 10^{15} GeV, which is at an equivalent temperature of 10^{28} kelvin. Such conditions are well beyond the present (or future capabilities) of man-made particle accelerators. The biggest machine today is the Large Hadron Collider at the CERN laboratory which gets us to 1.4×10^4 GeV. For those who hold that nothing should be accepted in science without it being experimentally verified, this is embarrassing. How is one ever conclusively to demonstrate that the forces do indeed merge into one another when this is expected to happen only at unattainable energies? All we can say is that at the energies that are reachable, the tendency is for the strengths of the forces to be converging towards each other.

But from here, the extrapolation to equal strength is a considerable one, and such unification remains something of an act of faith.

Can we ever be sure ≺
that GUT is correct
if we cannot
experimentally test it
at the appropriately
high energy?

If GUT is true, then the requisite energy conditions were presumably achieved in the early stages of the Big Bang. It is expected that, up to a Planck time after the instant of the Big Bang, these three forces, plus gravity, were united. Around the Planck time, gravity was the first to differentiate itself from the others by becoming weaker than the other three. The strong force then separated out at 10^{-36} seconds when the temperature had dropped to 10^{28} K. This was just before inflation kicked in between 10^{-36} seconds and 10^{-32} seconds.

Although for practical reasons it is unlikely that we shall ever be able to reproduce these high energy conditions in a laboratory, and thus test GUT directly, might there not be ways of testing the theory indirectly at lower energies? The answer is a tentative yes. GUT has the desirable feature of not only unifying the forces, but also of unifying quarks and leptons. Remember how we distinguished quarks from leptons by saying that leptons, unlike quarks, do not experience the strong force? But this apparently clear-cut distinction between them no longer holds once we recognise that the weak and strong forces are merely different manifestations of the same common force. Without that hard and fast distinction between them, quarks should be able to change into leptons, and vice versa. This leads to the prediction that protons should decay. They should transform, for example, into a positron (an antielectron) and a neutral pion. Attempts have been made to detect such decays. The most notable is with a detector called Super Kamiokande which is situated 1000 metres underground in Japan. It consists of a cylinder 39 m in diameter and 41 m high, containing 50,000 tons of very pure water. This holds 7.5×10^{33} protons. No example of a proton decaying has yet been seen. The lower limit on the lifetime is currently put as being of the order 10^{35} years. This is already much longer than a prediction of 10^{31} years based on a once-favoured version of GUT. Given that detectors significantly larger than the present one are unlikely ever to be built, there is concern that perhaps the actual lifetime is so long that we might never have the ability to measure it.

➣ *Shall we ever be able to verify proton decay?*

GUT also leads to the conclusion that there should be magnetic monopoles. Normally, of course, magnetic poles come in pairs—a north pole coupled to a south pole—but GUT suggests that there could be particles carrying but one pole, in much the same way as they carry one form of electric charge—either positive or negative. However, no such monopoles have been seen. This is slightly worrying because, although they might be very heavy and hence hard to produce in a particle accelerator, they ought to have been formed in the very early, extremely hot stages of the Big Bang, and thus one would think that we ought to be able to find them in nature as a remnant of that violent epoch. One reason why they are rare might have something to do with inflation. According to this suggestion, the monopoles are thought to have formed before the period of inflation, but then the hyperexpansion associated with inflation dispersed them far and wide and diluted them with all the new matter that was produced during inflation itself—the new matter that gave rise to the critical density and which contributed no more monopoles because the temperature had by then dropped too low. This might account for the fact that they have not been seen.

Thus we conclude that, attractive though the GUT hypothesis undoubtedly is, it is to date an unverified postulated extension of the Standard Model.

Why is there no evidence for the existence of magnetic monopoles? ◄

Are there any other concerns over the Standard Model? Yes. It requires 19 unknown parameters to be given values in order to make it work. There is nothing in the theory to set the values of the properties of the particles—the masses of the quarks and leptons and the strengths of the forces. These must be fed in by hand to fit the observed data. One would have thought that a really satisfactory theory would specify what these had to be. No one knows what rationale might lie behind the actual choice of values. Perhaps we shall never know. Another boundary of the knowable, perhaps.

Another concern is how the particles get their masses. A mechanism has been proposed as to how mass arises. One can think of it this way: In an electric field it can take an applied force to maintain the position of a charged particle. This is because we think of the field exerting a force on the particle tending to alter the particle's position. This force depends on the strength of the field at that position. So, in order for the particle to maintain its position, one has to apply a force to counter that arising from the electric field. Next consider an object (a ball, say) moving through a medium like water or treacle. This time it takes an applied force to maintain the speed. This is because of the viscosity force exerted on the ball by the medium. Finally, consider a particle accelerating through empty space. We find this time that it takes an applied force to maintain the acceleration. Why? The conventional approach is to say that the object has an inertial mass and that it takes an applied force to accelerate a massive body. But an alternative approach is to say that not only does it take an applied force to maintain a position in an electric field, and

> *Is it possible to account for the values of the parameters featured in the Standard Model?*

one to maintain a speed in a viscous medium, but that it also takes an applied force to maintain an acceleration in a vacuum. This is because the vacuum is filled with another type of field—one called the **Higgs field** (after the Scottish physicist, Peter Higgs who made the proposal in 1964). This field exerts a force resisting further acceleration. If there were to be such a force one would expect it to be mediated by an intermediate particle—just like the electric, weak, and colour forces. Thus, there is the proposal that there is such a new particle—one called the **Higgs particle**. It is expected to be very heavy; estimates put a lower limit on it of 120 m_p. Some theories point to the mass of the Higgs perhaps being several hundred times that of the proton. It is hoped that the recently commissioned accelerator at CERN, the Large Hadron Collider, might be able to generate enough energy to produce it. The Higgs mechanism would offer a way of understanding why an applied force is needed to accelerate a particle—or to put it another way—how we come to associate an 'inertial mass' with the particle. At the time of writing, no Higgs particle has been found. It is to be hoped that the LHC will discover it by the time you read this, or soon afterwards. But then again, it might remain elusive.

Is there a Higgs particle? ≺

Even with a Higgs particle, however, that still leaves wide open why the individual elementary particles have the particular masses they do. As we have seen, the masses of the quarks are wildly different from each other. The same is true of the leptons. The tau lepton, for example, has 3,520 times the mass of the electron lepton. Not only that, none of these masses is anything like what we might expect the masses of nature's fundamental building blocks of the universe to be. Recall how earlier we saw that there appeared to be a 'natural' unit of length and of time (the Planck values). In a similar way we can construct a 'natural' unit of mass out of the fundamental constants, h, c, and G—the **Planck mass**:

$$m_P = (hc/2\pi G)^{1/2} \approx 2.2 \times 10^{-8} \text{ kg}$$

It is perhaps puzzling that the masses of the fundamental particles are nothing like the Planck mass. For instance, the Planck mass, which weighs as much as a grain of dust, is 10^{19} times the mass of the proton.

➢ *How are we to account for the masses of the particles?*

One should reiterate that all this talk of the unification of forces leaves out of account the gravitational interaction. So far there has been no way of including it within the framework of Grand Unification. We shall say more about this problem later under the heading *quantum gravity*.

With regard to unification, there is one further type to consider. We have seen how we have two broad types of particle: on the one hand the quarks and leptons that constitute matter, and on the other the mediators of the forces—photon, gluons, W, Z, and Higgs. A theory named **supersymmetry** proposes that these two sets of particles are not as different from each other as one usually thinks. It is proposed that each has a partner of the other sort. The general idea is that if in a particular reaction all the particles—both the matter particles and the force-carrying particles—are replaced by their superpartner, the resulting behaviour remains the same. Thus the situation has a certain symmetry about it—hence the name supersymmetry. The reason for making such a proposal is rather technical. It has to do with the kind of quantum fluctuations that we saw earlier that gave rise to virtual pair production of particles. In the calculations these can very easily get out of hand and produce nonsense results. This in turn requires that the values of the various parameters be fine tuned to prevent trouble, all this seeming to be rather fortuitous and unlikely. The beauty of having supersymmetric partners is that the contributions from the superpartners largely cancel out the unwanted effects automatically, thus doing away with the need to fine tune the parameters. Furthermore the inclusion of superpartners is an integral feature of string theory which we deal with later.

The trouble is that no one has yet seen any of these supersymmetric partners. It has been suggested that this might be because they are all much heavier than the particles we know. Perhaps the LHC will be capable of verifying this. Unfortunately, from current theories, there is no way of predicting what the masses of the partners would be. Another worry is that even the simplest form of supersymmetry introduces 105 additional adjustable parameters to the 19 already incorporated into the Standard Model.

> *Does supersymmetry hold and, if so, why have we not as yet seen any of the supersymmetric partners?*

11

The quantum world

Quantum theory has come into our discussion a number of times already. To see what it is about, we begin by taking a look at the nature of light.

The study of light has had a chequered history. The Greeks thought that light was made up of particles. In the 17th century Christian Huygens, and independently Robert Hooke, advanced the view that it consisted of waves. At the same time, Isaac Newton championed the idea that it was made up of particles—or 'corpuscles' as he called them. Such was the fame of Newton in the wake of his successful theory of gravity, that the scientific community in general came down on the side of Newton.

It was Thomas Young, born in 1773 some 45 years after the death of Newton, who was able to demonstrate unambiguously the wave nature of light. Even before Young's time it was known that there was something odd about the appearance of shadows. Pass light through a large hole in a barrier and the edges of the shadow so cast on a screen on the far side look sharp. But if the size of the hole is reduced then the edges of the shadow become indistinct; the light appears to bend somewhat round the corner to overlap with what would otherwise have been the strict geometrical shadow. This is the phenomenon known as **diffraction**. One could attempt to explain such behaviour in terms of the particles of light scattering off the rim of the hole, but the observed pattern is found to be independent of the composition of the material making up the barrier—the material that is supposed to be doing the scattering. A further problem is that it is difficult to see why the scattering of the particles should increase the smaller the hole is made. A more plausible explanation is that the light is made up of waves. It was well known that waves spread out after passing through a hole—for instance, water waves passing through a gap in a sea wall.

Young's great contribution, in 1801, was that he placed two holes—well, narrow slits—very close to each other, such that the light diffracted through each slit overlapped on the screen on the far side. What he found was not two patches of light, one corresponding to each of the two slits, but a whole series of light and dark patches—what came to be called **interference** fringes (see Figure 19). It was found that there were regions where no light fell if both slits were open, but where there could be light if only one slit was open.

This was the death knell for the corpuscular theory of light. There was no way that light could scatter off the rim

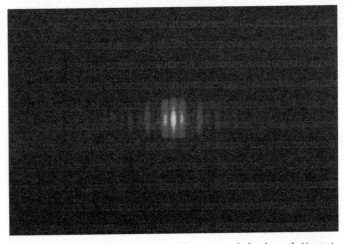

FIG 19 Interference fringes produced by passing light through Young's double slit arrangement.

of a hole and go to a particular location on the screen if only that hole were open, but it could no longer go to that region if the second hole—a hole it did *not* pass through—was also opened. The wave theory, on the other hand, had a simple explanation of the phenomenon. The waves being diffracted through the two slits were made up of a succession of peaks and troughs. When they overlapped on the screen, in some places the two beams were in step, with peaks arriving simultaneously from both slits, and troughs likewise arriving simultaneously, thus reinforcing each other (so-called **constructive interference**) and giving a bright patch of light. In other places on the screen the two beams were out of step, with peaks from one slit arriving at the same time as troughs from the other, thus leading to cancellation (**destructive**

interference) and darkness. All that was required was that the width of the slits and the separation of the two slits should be comparable to the wavelength of the light (i.e. the distance between successive peaks or troughs). If this condition is not satisfied then the two diffraction patches do not overlap and there can be no interference.

In 1821, Augustin Fresnel came up with the idea that the light waves were such that the vibrations were transverse to the line of motion of the light. Polarised light could then be explained in terms of the vibrations in one particular direction being cut out. The reflection and refraction of light could be explained in terms of waves, and the various colours of white light were recognised as being waves of differing wavelength. Finally, in 1864, as we saw earlier, James Clerk Maxwell, building on the experimental investigations of electricity and magnetism by Michael Faraday, predicted that there should be electromagnetic waves, and that these should travel at a speed which turned out to be the speed of light. This showed that light, along with infrared, radio, microwave, ultraviolet, X-rays, and gamma rays were all forms of electromagnetic waves, differing only in their wavelength and frequency. Thus the issue was settled once and for all: light is a wave.

But wait. All we have talked about so far is how light travels about—through space, through glass, through holes in barriers, through polaroid spectacles, etc. But what happens when it reaches its destination—when it collides with something and exchanges energy and momentum? With it being electromagnetic waves, i.e. a succession of tiny vibrating electric and magnetic forces, one might anticipate that when it interacts with, say, the screen lying beyond the barrier with the holes, it would gently set the atomic electrons of the screen vibrating. These vibrations would gradually build up in amplitude—rather like a child's swing being put in motion by continued regular pushing. Eventually some of the electrons would loosen their grip on their parent nucleus and drift off.

What actually happens is quite startling. It is found that as soon as light begins to fall on a metal plate, electrons are immediately ejected. Far from all the electrons being gently set moving about within their atom, all the energy of the light becomes concentrated on but a few of the electrons. These are immediately and forcibly ejected from the plate, leaving the vast majority of electrons completely unaffected. This phenomenon is known as the **photoelectric effect**. It was Albert Einstein who came up with the revolutionary idea that—contrary to received opinion—light under these circumstances was behaving, not as a wave, but as a stream of particles! These particles of light hit only certain of the electrons, ejecting them in the process, leaving the other electrons undisturbed. It was as though the metal plate was being subjected to a hail of gunfire. These particles of light are the photons to which we have made reference from time to time.

It was found that, like ordinary particles, photons carried energy and momentum. The energy, E, was given by the expression

$$E = hf$$

where h is Planck's constant, and f is the frequency of the light (related to the wavelength, λ, and the speed of light, c, by $f = c/\lambda$).

The momentum, p, of the photon is given by

$$p = hf/c$$

The fact that light possesses energy is a familiar one; we are, for example, used to the idea of surfaces exposed to sunlight heating up. But momentum too? Yes, this can be demonstrated by shining light onto a paddle wheel mounted onto an almost frictionless bearing. Shining the light on one side of the paddle will set the wheel rotating. In other words, the paddle is being 'pushed' by the light, i.e. momentum is being transferred from the light to the wheel. On a grander scale we can observe the same phenomenon in the behaviour of comet's tails. It is widely believed that as the comet head rushes through space, its tail stretches out behind marking where it has been. But this is incorrect. The tail does not lie along the direction of motion of the comet, but instead stretches out in a direction away from the Sun. It is the pressure exerted by the Sun's light that is pushing the tail away from it.

Examining the expression for the energy of the photon we see that it is directly proportional to the frequency of the light. This can be checked by measuring the energy of ejection of the electrons from the metal plate as the frequency of the light is varied. It takes a certain amount of energy—the threshold energy—to extract an electron away from the plate. So, below a certain frequency (or photon energy) no electrons are emitted. Above that threshold, the ejected electron takes up the energy of the photon minus the amount needed to extract it from the plate.

Given the prevalence of the wave theory of light at the time, it was a bold move on Einstein's part to resurrect the particle idea once more. This he did in 1905—the very same year as he proposed his special theory of relativity. In this way, we are presented with a situation where certain experiments—those concerned with how light gets from A to B—have to be interpreted according to the wave theory of light, while other experiments—those concerned with how light interacts when it is emitted at A and how it interacts when it reaches its destination at B—have to be interpreted according to the particle theory. This somewhat paradoxical outcome is known as **wave—particle duality**.

Nor is this a schizophrenic behaviour characteristic solely of light. Take the electron. So far we have been thinking of the electron solely in terms of it being a point-like particle. That is certainly how it appears to be when it interacts—for example, when it collides with another electron, or a proton, say. Passing through holes in a barrier it appears to go straight on, giving a sharp shadow—the kind of shadow one gets if one squirts a paint spray through a hole—the paint spray being made up of tiny droplets (or 'particles') of paint. But be careful! We get quite sharp shadows if we pass light through a hole which has dimensions large compared to the light's wavelength. Diffraction only becomes apparent when light passes through a hole whose size is comparable to that of the wavelength. What if the electron had wavelike characteristics but that these do not show up readily because the wavelength is tiny? We saw how the momentum of a photon was related to the frequency of the light by $p = hf/c$, or writing this in terms of the wavelength: $p = h/\lambda$. Suppose the same formula applied to the electron. Immediately we are able to conclude that, because the momentum of the typical electron is much, much greater than that expected of a photon, the supposed wavelength associated with the electron would be absolutely tiny. Thus to be able to see diffraction and interference effects, the sizes of the holes involved would have to be correspondingly smaller.

In 1927, Davisson and Germer were able to demonstrate diffraction of electrons using a nickel crystal. Because the atoms that make up a crystal are arranged in a periodic and orderly manner, the crystal acts like a diffraction grating. A diffraction grating consists of closely spaced parallel lines. In the case of the crystal, the planes containing the atoms act like the parallel lines, and the gaps between the planes fulfilling the same function as the gaps between the lines of the grating. Because the planes containing the atoms lie so close to each other, the gaps are very narrow—narrow enough to give rise to noticeable diffraction patterns even though the wavelength of the electrons is so small. Indeed, these days by accelerating electrons to known energy and hence known momentum, and using the above formula connecting momentum to wavelength, it has become standard practice to use such beams and the observed configuration of the resulting interference patterns to examine the structure of the crystal. In other words, electron waves have become a standard laboratory tool.

In fact, experiments have now been carried out which clearly show that beams of electrons get diffracted when passing the edges of objects or going through holes. As of 1961, even the Young's two-slit experiment has been successfully carried out for electrons, showing interference fringes.

Thus, an electron beam is not only characterised by properties that we normally associate with particles—energy and momentum—but also a property associated with waves—a wavelength. This wavelength is called the **de Broglie wavelength** of the electron, named after Louis de Broglie, the French physicist who first drew attention to the wave characteristics of the electron. The wavelength, λ, is given in terms of the electron's momentum, p, and Planck's constant, h, by the expression:

$$\lambda = h/p$$

De Broglie did not come across the wave nature of the electron through diffraction and interference experiments—these had yet to be carried out. Rather, he arrived at this conclusion through a study of the properties of atoms. The favoured model of the atom at that time was that which had been devised in 1915 by Neils Bohr. It was known that atoms consisted of a nucleus surrounded by electrons, but that raised the question as to what prevented the electrons from being attracted into the nucleus by their opposite electric charges. Bohr had suggested that the electrons were orbiting the nucleus in much the same way that planets resist being gravitationally pulled into the Sun by orbiting it. That seemed straightforward enough, except that Bohr further postulated that the electrons were capable of following only certain allowed paths, each path being characterised by a particular energy for the electron. This suggestion derived from the characteristics of the spectra of light emitted by atoms. If atoms are heated up, the electrons gain energy and move to higher, more energetic orbits. They then subsequently cascade down to the lower orbits with the emission of light. It is found that light emitted by atoms has only certain discrete wavelengths, the actual values of the allowed wavelengths being characteristic of the particular atom. Heat up sodium for instance and one gets a yellow light (that of the familiar sodium street lights). Neon emits red. It is the pattern of the wavelengths emitted by unknown substances that allows us to identify the atoms doing the emitting. The discrete nature of the wavelengths points to the accompanying photons having discrete allowed values for their energy (recall $E = hf = hc/\lambda$). Their energy derives from the difference in energy between the initial and final states of the transition between one orbit and the other. Bohr proposed that the orbiting electrons could possess only certain allowed values for their energy, and hence only certain orbits were allowed. In order to emit a photon, the electron had to perform a quantum jump from one allowed state to another allowed state—hence the discrete energy differences which lead to the discrete wavelengths for the emitted light.

The idea that only certain orbits are allowed laid to rest a worry that had afflicted the planetary model of the atom. Electrons, unlike planets, are electrically charged. It is known that charged particles when undergoing acceleration radiate energy. Particle accelerators that depend on electrons going round a closed circle lose energy constantly due to the centrifugal acceleration. That's why the most powerful electron accelerators are straight, like the two-mile linear accelerator at Stanford, California. So the question arises as to why an electron orbiting the nucleus in an atom doesn't gradually lose energy and spiral into the nucleus. Bohr's answer was that it cannot gradually and continuously lose energy because that would take the electron into a nearby orbit—one that was not a permitted orbit.

All of this, of course, begs the question as to why only certain orbits are allowed. This is where de Broglie enters the story. In 1923, he made the suggestion that the electron was wavelike—as audacious a suggestion as Einstein's, concerning the particle-like nature of light. De Broglie argued that being a wave it would have to fit exactly into the orbit with an integral number of wavelengths. If not, the wave train being wrapped round the orbit would repeatedly get out of step with itself, with peaks cancelling out troughs, resulting in nothing. Only with an exact number of wavelengths around the orbit would all peaks reinforce themselves and the same with the troughs. Thus only certain wavelengths were permitted, and consequently only certain electron energies.

Nowadays we know that the Bohr model is too simplistic. When one solves the relevant equations it turns out that the electrons are not neatly distributed between clean-cut obits; they are spread out in various patterns. But although the notion of orbits is wrong, the essential idea is correct, namely that because we are dealing with confined waves, those waves will be characterised by discrete values for their wavelength and frequency. As with waves confined between the end points of a guitar string, only certain vibration patterns are allowed—which in the case of a guitar means only certain discrete notes can be sounded (the fundamental and its harmonics), while in the case of an atom, only certain discrete energy states are open to the electron.

Add to this the more modern observations of electron diffraction and interference, and one concludes that the wave nature of the electron is as well established as the earlier particle characteristics.

And it is not only photons and electrons that are subject to wave–particle duality; other types of matter are similarly affected. Beams of protons, heavier nuclei, complete atoms and complete molecules have been subsequently shown to have the same characteristics. Such particles being heavier tend to have more momentum and, as can be seen from the de Broglie formula given above, they have smaller wavelengths.

The diffraction and interference effects they give rise to are correspondingly more difficult to detect, but they are there for all types of matter.

One might also add in parentheses, that the above considerations also apply to your own body. As you walk, you will be characterised by a de Broglie wavelength, the value of which will depend on your momentum in the same manner as the wavelengths for light, electrons, and other matter particles. The faster you walk the smaller the wavelength. As you pass through a doorway (a 'hole'), you will be diffracted—but not by very much. No drunkard staggering out of a pub onto the street is likely to be able to convince a police officer by claiming he was diffracted!

So, summarising to date, we find that the contents of the world (matter and light) sometimes behave like a wave, and sometimes as a particle or **quantum**—a quantum having a discrete quantity of energy and momentum. It all sounds very confusing. After all, how can something be both a spread-out wave with its peaks and troughs and, at the same time, a tiny localised particle. This is the so-called **wave–particle paradox**.

Nevertheless, there is a certain sense of order to it all. Whether we are dealing with a wave or a quantum depends on the type of question being asked. If we are asking how does something travel through space—where is it going to go—then we call on the behaviour characteristic of waves. If on the other hand we are asking a question to do with how something interacts—involving the transfer of energy and momentum—then we have to switch over to the behaviour of particles. Either we are talking about something *passing between* A and B, or alternatively we are talking about what it is doing *at* A or B. It is one or the other; it cannot be both. Hence, there is no call to use the wave and the particle descriptions at the same time. Thus in practical terms there is no ambiguity.

But, you might be asking, what kind of waves are we talking about? After all we are used to waves being the movement of some sort of medium—waves on water or sound waves in air, for instance. Or in the case of electro-magnetic waves, which do not rely on a medium as such, they are themselves very physical things, being made up of electric and magnetic vibrations.

The clue as to the nature of this new kind of wave is to be found in an experiment carried out in 1909 by Geoffrey Ingram Taylor, four years after Einstein's suggestion of the photon-like nature of light. He repeated the Young's double slit experiment but with a very weak source of light—as he put it, 'the equivalent of a candle burning at a distance slightly exceeding that of a mile'. This meant that the amount of light energy passing through the system at any one time was no more than that expected of a single photon. That being the case, what would one expect to see on the screen? Would the photon still arrive at just one point on the screen, or would it smear itself out to form the interference fringe pattern?

In fact, it was found to arrive at just one point. So does that mean the wave nature of light is, for some reason, discounted under these conditions? No. We have to be patient. A second photon passes through and also lands up at a particular point on the screen, then a third, and a fourth, etc. After a while we notice that there are regions on the screen that never receive photons and others where they arrive frequently. Over time, the interference pattern is progressively built up with its characteristic light and dark fringes (Figure 20). Thus we find that the wave-like nature of the light was present all along. It was working behind the scenes, so to speak, governing the probabilities of where the individual photons were likely to go. For this reason, they are sometimes referred to as **probability waves**.

We said earlier that the wave description was the appropriate one for describing how light or electrons move through space—and hence where they were going to go. Now we recognise that, more strictly speaking, the wave nature determines where the light will *probably* go—it provides the means for calculating the relative probabilities for finding the photon at various locations. And not just photons. In the 1970s, it was experimentally demonstrated that electrons being passed singly through a double slit arrangement behaved in the same way. Thus we conclude that, with the wave nature being associated with radiation of all sorts—whether beams of light, electrons, protons, atoms, or molecules—they are all to be regarded as probability waves.

The idea that all we can talk about are probabilities is not very satisfactory. One is tempted to think that perhaps light, for instance, is actually at all times made up of quanta and the position where we find a particular quantum on the

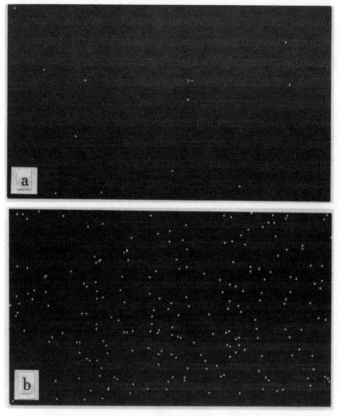

FIG 20 Photons passed individually through a double slit arrangement arrive at discrete points on the screen, but nevertheless, over time, they build up the familiar interference pattern.

screen is determined by how it scatters off the rim of the particular slit it passes through. All we have to do is examine sufficiently carefully the details of how the photon bounces off the rim. But this will not do. We have already pointed out

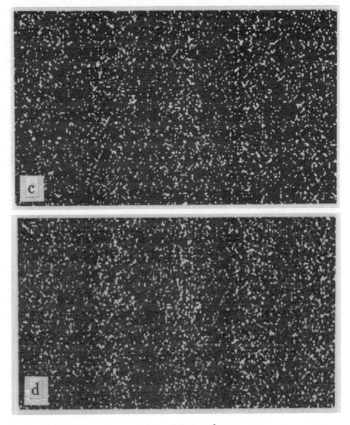

FIG 20 Continued.

that the diffraction pattern does not depend on the material out of which the barrier is made—the material involved in the supposed deflection of the photon. Moreover, it is impossible to understand how the opening of a second slit (one the photon does not go through) affects how it scatters

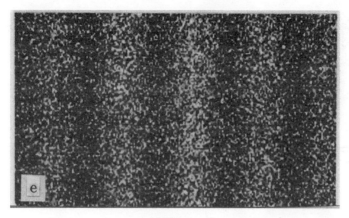

FIG 20 Continued.

from the rim of the slit it does go through. No, we are faced with the fact that we have to deal only in probabilities.

It was in 1926 that Erwin Schrödinger developed the mathematics for handling the probability waves. He produced what we now call the **Schrödinger wave equation**. The solutions to this equation are called **wave functions**. Solving the wave equation for the double slit experiment produces a wave function that determines the probability distribution of the light on the screen. Solving the equation for the case of an electron bound to a proton—a hydrogen atom—produces a wave function that tells us the relative probabilities for finding the electron in different locations within the atom (Figure 21).

The spread-out nature of the wave function can lead to some strange consequences. For instance, in some respects, the neutrons and protons in a nucleus behave as though they

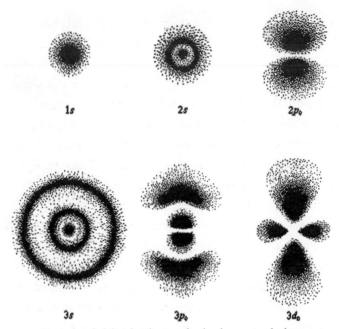

1s 2s 2p₀

3s 3p₀ 3d₀

FIG 21 Typical probability distributions for the electron in a hydrogen atom.

are clumped together like helium nuclei, each consisting of two protons and two neutrons. These helium nuclei—or as they are commonly called **alpha particles**—move around within the nucleus, and in doing so have a wave nature. The kinetic energy of the alpha particles is not great enough to overcome the strong force of attraction to the other constituents of the nucleus, so one would think that they could never escape. But occasionally they do. This is because their wave function, to a very tiny degree, extends beyond the confines of the overall nucleus to which the alpha particle belongs. This means that there is a finite probability of finding the alpha particle outside the range of the attractive force—although according to classical ideas of energy there is no way it could find itself there. And once the alpha particle finds itself there, beyond the range of the attractive nuclear force, all it now experiences is the long-range electric repulsion of the protons remaining inside the nucleus, and so is propelled away. The nucleus thereby loses an alpha particle—it undergoes radioactive decay. This well-known phenomenon is known as **alpha particle decay.**

As an aside, perhaps I should mention that all objects have a wave function determining where we are likely to find them—a grain of sugar, for example. The wave function of such a grain will extend to a very tiny extent beyond the sealed package containing the sugar—just as the wave function of the alpha particle extends a little way beyond the nucleus containing it. But no, this is not the explanation of why one sometimes finds a trace of sugar on the kitchen shelf beside the bag! The leakage of the wave function in this case is far, far too small to produce any noticeable effect.

The introduction of the concept of probability goes against the 'common sense' notion of determinism. According to Laplace, if he could be given the data on all the particles—meaning their positions and velocities (or momentum)—he could predict the future of the universe. Fair enough, but how is one to gather all the relevant initial data?

Let us begin with just a single electron. First, consider how one would measure its velocity. We need to take a look at it.

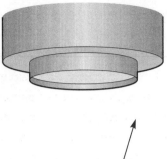

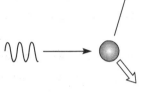

FIG 22 A hypothetical gamma ray microscope for trying to measure the position and momentum of an electron.

We imagine shining a light onto it, and examining what happens under a microscope (Figure 22). But in doing so we know that the light will interact as a photon carrying energy and momentum. It knocks the electron being observed and in the process disturbs it. So, whatever the velocity or momentum of the electron was before the measurement, there is uncertainty about its value after it has been altered in the collision.

The answer to this is to use low frequency radiation, i.e. photons of little momentum. In the limit of very low frequency, they will not disturb the electron at all, so one could determine the electron's momentum to any accuracy one likes. But there is a down side to this. Such light has a long wavelength. It is spread out more than light of a compact short wavelength. This in turn means that we cannot be sure *where* exactly the interaction took place. So we might know the momentum of the electron but not the position.

To get the position precisely we need very short wavelength radiation—gamma radiation. In the limit of almost zero wavelength we can achieve perfect precision over the determination of the position of the interaction, but as we have seen, that tends to knock the electron flying and we lose information on its momentum.

Thus, by using alternative types of light we can gain perfect precision of either the momentum or the position of the electron, but not both at the same time. It becomes a trade-off. More precision with one parameter goes hand in hand with a loss of precision with the other, but in order to determine future behaviour we need both.

This hypothetical 'gamma ray microscope experiment'—as it came to be called—was discussed in 1927 by Werner Heisenberg and led to the formulation of the **uncertainty relation** which bears his name:

$$\Delta x . \Delta p \approx h/2\pi$$

where Δx and Δp are the uncertainties in position and momentum respectively. As noted before, Planck's constant, h, is very, very small, so the uncertainties we are talking about are small. They become significant only on the atomic and subatomic scale. From this relation we see that in order to reduce one of these uncertainties to zero, the other has to be infinite. In other words, perfect precision for either parameter has to be bought at the price of losing all information about the other. Either that or one opts for a compromise: some limited information about both. It is because of the uncertainty relation that we cannot, even in principle, gain all the information required to predict future behaviour exactly.

An alternative version of the uncertainty relation is

$$\Delta E.\Delta t \approx h/2\pi$$

This arose earlier in our discussion of the nature of space. There we introduced the idea of pairs of virtual particles coming into existence (page 61). This process takes energy. It is the uncertainty in energy, ΔE, that allows the virtual particles to borrow some for a short time, Δt, and materialise. The process is governed by this uncertainty relation which specifies how long the energy fluctuation can last before being paid back. The heavier the particles of the pair, the more energy they will need, and hence the shorter the period of time they can exist.

The uncertainty relation applies to every measurement of energy. ΔE is the unavoidable uncertainty in the measurement arising from an uncertainty, Δt, in the time that the measurement is made. For example, suppose an atomic electron is put into a higher energy state, from which it subsequently makes a transition to a lower energy state. The fact that it was in the higher unstable state for only a limited time means that any measurement of the energy of that particle in the higher state must refer to its energy during that limited time. This imposes an upper limit on the uncertainty, Δt, that can be associated with the estimate as to when the measurement was made. That in turn implies, through the above uncertainty relation, that there will be a lower limit on the uncertainty, ΔE, in the estimate of the energy of the electron while in that state. We cannot know its energy with complete precision ($\Delta E = 0$) unless $\Delta t = \infty$, i.e. there would have to be an infinite time available for making the measurement.

But you might ask, is there no way of getting round the uncertainty relations in order to obtain all the information we want from the electron? In the 1930s, Einstein was prominent among those physicists who tried devising ever more sophisticated and ingenious strategies aimed at gaining perfect knowledge of both position and momentum. Neils Bohr played a major part in refuting such attempts. Gradually, and it has to be said, grudgingly, it came to be recognised that here one was indeed up against a barrier of the knowable. Whatever the precise position and momentum of the electron might be, we were never going to gain access to both.

We have seen that if something is interacting we are to treat it as a quantum (or particle); if on the other hand one is asking about its whereabouts—how it moves through space—then we are to treat it as a wave. It cannot be both travelling towards it destination at the same time as it is interacting at its destination. Hence there is no ambiguity as to which description to use. Thus the problem of wave–particle duality is solved.

Except, of course, that this is but a recipe of what to use when. It does not address the fundamental problem of what exactly *is* an electron. Suppose, for example, there is an isolated electron out there in space and we are not asking any particular question about where it is or what interaction it might be undergoing. Under those circumstances, what is the appropriate description of the electron? Are we to envisage it as a wave or as a particle? Is it really a wave that sometimes gets a bit 'gritty', or is it really a particle that just happens to get a bit 'wavy' at times?

Although all physicists are agreed as to how to *apply* quantum mechanics to solve problems, using the Schrödinger wave equation, there has always been, and continues to be controversy as to how to *interpret* quantum theory. In other words, what does it all mean? What actually is going on? There is no consensus. In what follows, I will try to summarise the various thoughts that people have had on the subject. It will then be up to you to make up your own mind as to which sounds the more reasonable.

In the first place, we have to ask what significance should be given to the wave function. According to one school of thought, the wave function applies merely to an ensemble of measurements. In the case of the two-slit interference experiment, the wave function determines the spread of light on the screen—how an ensemble of photons will be distributed over the screen. As such, it leaves open the question as to whether a detailed study of individual events might reveal why a particular photon goes to its exact location on the screen. The wave function is just a broad-brush approach dealing in average behaviours. This is an interpretation favoured by those who, like Einstein, regarded quantum theory as incomplete. The theory is true as far as it goes, but it falls short of providing a full explanation of reality.

Bohr did not go along with this. He was opposed to any idea that the photon or electron was actually behaving much like a classical particle, possessing at all times both an exact position and an exact momentum, moving through space in a deterministic manner and that it was just our clumsy measuring techniques that denied us complete knowledge and reduced us to statements of a statistical nature relating to an ensemble of such particles. As far as he was concerned, parameters such as position and momentum were mutually exclusive concepts—they were complementary. They could not both take on exact values at the same time. Thus, although the wave function certainly does describe how an ensemble of particles would behave, it additionally and more importantly, describes the individual event. The concept of probability is inherent within the arrival of the individual quantum on the screen. This view came to be known as the **Copenhagen interpretation**—so named after the city where Bohr worked.

The term Copenhagen interpretation in fact covers a number of variants, and it becomes important to distinguish between them. Their common factor is that they all hold that the wave function, in some sense, applies to each and every individual situation and not just to the averaged out behaviour of many similar situations. This means that each and every event incorporates an element of probability. Bohr insisted that each event is intrinsically uncertain, thus ruling out any possibility of being able to determine exactly where the photon will arrive on the screen.

We said that the wave function applies to each event 'in some sense', but how exactly does it apply to the individual situation?

One variant is to say that the wave function is a description of the actual state of the system under investigation at each point in time. Suppose, for example, we have an electron passing through a hole. At the instant of passage we know its position, and we can know it as precisely as we like by making the hole small. Thereafter we do not know for certain where it will next be found. The wave function spreads out— the phenomenon that we have been calling diffraction. For all we know, the electron might subsequently be found any-where on the screen—or at least anywhere within the scope of the diffraction pattern. In other words, immediately before the electron arrives at the screen, the wave function—and hence the electron itself—is considered to be spread out over a large area. But then, at the instant of its arrival on the screen, the situation is transformed. We again know the exact point at which the electron is located. The wave function has therefore been instantaneously transformed from being spread out to being localised at the point of arrival.

This raises a difficulty. It requires an instantaneous trans-formation of the wave function—the so-called **collapse of**

the wave function. The furthest reaches of the wave function have to acquire knowledge of the localisation of the electron, and this is seemingly done instantaneously. This appears to violate relativity which says that no signal can travel faster than the speed of light. It was for this reason that Einstein preferred the ensemble interpretation of the wave function.

The difficulties encountered by the claim that the wave function is a description of the physical system were thrown into high relief in 1935 when Erwin Schrödinger proposed his famous (or infamous) **Schrödinger's cat experiment**. The idea is that a cat is placed in a steel chamber, along with a device containing a vial of prussic acid. The chamber contains a very small amount of a radioactive substance. If even a single atom of the substance decays during the test period, a Geiger counter will detect the emitted alpha particle and immediately send an electrical impulse to activate a hammer which, in turn, breaks the vial, releases the poison, and kills the cat. Once the lid of the chamber is closed, the observer cannot know whether or not an atom of the substance has decayed and, consequently, cannot know whether the vial has been broken, the poison released, and the cat killed. The cat's wave function will therefore be a superposition of those corresponding to the cat being alive and the cat being dead. From a state where the cat is 100% alive at the instant the lid of the chamber is closed (observation 1), the mixture of live cat and dead cat progressively changes with time as it becomes more and more likely that the radioactive substance will have decayed. Thus, argued Schrödinger, the cat must be a mixture of being alive and being dead, until the lid of the chamber is subsequently lifted (observation 2), whereupon the cat instantaneously reverts to being either 100% alive again, or 100% dead.

Essentially it is the same problem we had over the collapse of the wave function of the electron passing through the hole and arriving on the screen. Schrödinger's version of the paradox, however, captures the imagination more vividly as it involves a familiar macroscopic object (a cat) rather than some subatomic entity that might, for all we know, do bizarre things like suddenly transforming from being smeared out to being localised. How can a cat be both alive and dead? And how can the lifting of a lid and taking a peek instantaneously transform the cat from a live/dead mixture to one that is either wholly alive or wholly dead?

The Schrödinger cat situation obviously poses problems for the variant of the Copenhagen interpretation that holds that the wave function is describing the state of the cat. Whenever the Copenhagen interpretation is attacked, it is usually this variant one has in mind. But there is another version. Although it was sometimes hard to understand what Bohr was driving at, and occasionally he could be seemingly inconsistent, it appears that it is this second variant that was closer to Bohr's own thinking. This version holds that the wave function does not so much describe the state of the physical system as the state of our *knowledge* of the system. On closing the lid of the chamber we know that the cat is 100% alive. Thereafter, the odds on finding the cat still alive when next we look into the chamber progressively change. The longer we leave it, the more likely we are to find the cat dead next time we look. The wave function is telling us how those odds are changing. But—and this is the crucial point—it is not telling us what the state of the cat is at any of these times. It merely tells us that, if we were to open the lid at any point in time, these are the odds on what you are likely to find. Prior to opening the lid one only knows the relative probabilities that we shall find the cat dead or alive; on opening the lid, we know what the situation actually is. Moreover we gain this knowledge instantaneously; there is no question of signals travelling faster than light. For this version of the Copenhagen interpretation, Schrödinger's cat poses no problem.

In this regard, I think it helpful to think of a slightly different version of the experiment. Forget about the cat. Instead, our experimenter, Physicist 1, places a fellow quantum physicist (Physicist 2)—someone who is just as capable of working out wave functions as Physicist 1—in the chamber. Let us say that the experiment has been running for a while to the point where the outside observer, Physicist 1, calculates that there is a 50% chance that the poison has been released and consequently Physicist 2 is dead by now, and a 50% chance that he is still alive. How does Physicist 2 see the situation? He will know for certain whether he is alive or dead. How? Because he is constantly making observations on the state of the vial of poison (and in any case knows from direct experience whether he is still alive). These observations yield a wave function for the poison, which carries the implication that he is 100% alive—assuming that he is indeed alive at this stage. Thus, Physicist 2 has his 100% alive type of wave function, while at exactly the same time, Physicist 1 on the outside, who only has his initial observation of the vial of poison at the time the lid was closed to go on, has a wave function yielding odds of 50/50 that Physicist 2 is now dead. So, which of these two wave functions is actually describing the state of Physicist 2 at that instant? The answer is neither. Both wave functions are doing nothing more than describing the state of *knowledge* of our two observers, and the functions differ from each other because the two observers are not working from the same observational data.

This, you might think, all sounds very reasonable, but if that is so, why aren't all physicists happy with this variant of the Copenhagen stance? The reason is not so much what this version says, but what it is *not* saying. All it ever talks about are observations—information about the results of instantaneous quantum observations. But what is happening *in between* the observations? Here the Copenhagen interpretation is silent—completely silent. And quite deliberately so. Bohr held that all we can meaningfully talk about are our observations of the world—our interactions with the world. What happens in between those interactions is completely unknown—and will for all time remain unknown. He held that all the words we use in physics—wave, particle,

wavelength, position, energy, momentum, electron, etc.—
are all words that are part of the vocabulary of accounting
for our interactions with the world. It is a *misuse of language* to
try to apply that sort of language to a description of the
world outside the context of a particular type of observation.
Should we ever take that invalid step, we get embroiled in
paradox. For example, we find ourselves asking how something can be both a spread-out wave and at the same time a
tiny localised particle, or we are questioning by what means
information can travel instantaneously across space in
apparent defiance of relativity theory.

It was actually Max Born who first advocated the idea that
the wave function was simply a description of our knowledge of the physical system rather than a description of the
system itself. But when he first put forward this notion he
had in mind that the physical system was, behind the scenes
so to speak, behaving in a classical manner. Bohr, as we have
now seen, went further and held that all talk about
what might be going on between observations was to be
discounted.

This is a severe restriction. To see how severe, let us revisit
the Heisenberg gamma ray microscope experiment. Recall the
procedure for trying to measure both the position and the
momentum of a particle by shining electromagnetic radiation of different wavelengths on it. This presupposed that
there actually was an electron out there, and that it did
possess an exact position and momentum—in other words
it was a classical particle. It was merely the limitations of our

measuring technique (knocking the poor electron flying with our photons) that was responsible for the uncertainty. That was the interpretation that Heisenberg himself put on it when he first proposed this hypothetical experiment. Bohr argued against him, saying that it was not right to think of the electron as really a particle, a compact object possessing an exact position and momentum. That would be to ignore the equally compelling evidence for a wave interpretation of the nature of the electron. The notions of particle and wave must both be given equal weight. And because a wave does not have an exact position, it is wrong to think of the electron itself sitting out there with an exact position, or an exact momentum, and all we have to do is try to find ways of getting round the uncertainty principle to discover what those values are. As mentioned before, many ingenious proposals have been put forward for experiments aimed at somehow getting round the uncertainty relation and gaining exact information about both position and momentum, at least in principle. All have failed. And if one cannot measure (i.e. observe) both an exact position and an exact momentum, then Bohr maintained that one is not justified in claiming that the electron actually *has* such exact values for these two parameters. Heisenberg himself was quickly won over by Bohr's arguments and was later to declare:

> It is possible to ask whether there is still concealed behind the statistical universe of perception a 'true' universe in which the law of causality would be valid. But such speculation seems to us to be without value and meaningless,

for physics must confine itself to the description of the relationship between perceptions.

W. Heisenberg, Zeitschrift für Physik, 43, 197 (1927)

We can summarise the argument as follows: It was widely believed that the job of science was to describe the world. In order to do this, we had to have a look at it to see what kind of world we were dealing with, i.e. we had to experiment on it, we had to observe it. But, having made our observations, what we wrote down in the physics books was a description of the way that the world was regardless of whether anyone happened still to be looking at it. Indeed, recall the opening words of Chapter 2: 'The task of science is to explain the world—what it consists of, how it operates, and how it came to be the way it is.' But according to Bohr, what science does is actually nothing of the sort. What we write down in the textbooks is not a description of the world at all but is effectively a description of us looking at the world—interacting with the world. If we try to go beyond the interaction in an attempt to describe the world *itself* we get bogged down in those paradoxes that arise out of the misuse of language. By contrast, there are no paradoxes when language is used correctly to describe solely our observations.

That is what most people regard as being the standard Copenhagen interpretation, i.e. we can speak meaningfully only of our observations of the world; we can say nothing of what might be happening in between the observations—out there in the world-in-itself. And there are a couple of variants of this which make more radical claims.

The first of these holds that we can say nothing about the world in between observations because there *is* nothing in between the observations. Physical reality consists simply of what are known as 'observations'. Quantum observations are discontinuous. For instance, one observation might be the measurement of the exact position of an electron. The second observation is a repeated position measurement some time later. In between these two instantaneous measurements, no observation is taking place, and hence, according to this extreme interpretation, nothing exists.

It is not difficult to take issue with this view. We know, in particular, that two such observational events are correlated. Nothing can travel faster than light, so this sets a limit on how far from the first measured position the second measured position can be. If absolutely nothing at all exists between those two instants, it is hard—indeed impossible—to see how the information about the first observation is to be retained and later correlated with the second. It thus seems indefensible to hold that the world-in-itself does not exist apart from observations of it. And yet this is a view that is still sometimes propounded.

Does the world-in-itself exist between our discontinuous observations? ◄

A second extreme version of the Copenhagen interpretation concerns the exact meaning of the term 'observation'. In normal use the word 'observation' implies an observer—a conscious observer. Thus there arises the notion that somehow consciousness comes into the reckoning—it is the thoughts of the observer that are responsible for producing the measured values of the parameters. Indeed, some would claim that the world would not exist if we were not thinking about it. But this view is widely dismissed as unwarranted. When we speak of making an observation in the context of quantum mechanics, we have in mind a physical apparatus recording the value of the position or momentum of an electron, say. Whether a conscious observer then takes note of the result of that measurement is neither here nor there. It is what is happening in the physical world that matters.

Not everyone, by any means, went along with Bohr when he proposed the idea that all we can meaningfully talk about are observations of the world rather than what might be happening in the world-in-itself in the gaps between the observations. The opposing camp had some prominent members: Einstein, Schrödinger, and Dirac, for instance. David Bohm proposed a 'hidden variables' theory. According to him quantum theory offered merely a statistical description of the behaviour of an ensemble of particles. But in truth there was an underlying reality that behaved deterministically, not only at the macroscopic everyday level, but also at the subatomic level. An electron, for example, was to be regarded as a localised particle—one that passed through just one slit of a double slit arrangement.

But, following up a suggestion of de Broglie, Bohm held that this particle was accompanied by 'a pilot wave' responsible for guiding the electron. It was the wave that sensed whether or not the second slit was open or closed, and guided its electron to the screen accordingly. So, although the particle could be regarded as localised, the wave was not, and moreover this wave was able to transmit information on the second slit instantaneously to the particle. Though Bohm's theory was much discussed for a while, it did not catch on. It was generally regarded as too contrived. Einstein did not like it. He moved away from a concern about determinism, and instead concentrated on emphasising the need to retain the notion of the reality of the world-in-itself. Regardless of what we might or might not be able to say about the world when it is not being observed, there is nevertheless a real world out there whether or not it is being observed.

While on the subject of different interpretations of quantum mechanics, I suppose we ought to mention the **many worlds** theory put forward by Hugh Everett in 1957. According to Everett, what happens when a photon passes through the two slit system is that the world splits up into myriad different worlds where every possible outcome of the observation is realised in one or other of those worlds. And this includes the observer. When the electron arrives on the screen, the observer splits up into many observers each of which sees the electron in different locations. Thus no choice has to be made as to where the electron goes—every possibility is realised in one or other these many worlds. All these spawned worlds are independent of each other. No sooner are they formed than they go their own separate ways, there being no possibility of one version of the observer having any contact with any other version. According to this interpretation, the wave function is describing this whole ensemble of outcomes in these newly splintered-off worlds. And what happened when the first electron arrived at the screen, is set to repeat again when the second arrives. Each of the infinite number of worlds generated by the arrival of the first electron now splits up into an infinite number of other worlds. And the same for the third electron—and indeed for everything happening in the world: splitting and splitting and splitting. Infinity compounded on infinity compounded on infinity. As such it might be regarded as even more extravagant than the multiverse idea invoked to explain the anthropic principle. That just involved infinity; this Everett idea compounds infinities on infinities.

This many-worlds interpretation is surprisingly popular among some physicists. I say 'surprising' because I myself find the whole idea somewhat devoid of explanatory value—to put it politely! There is certainly no way of proving or disproving the existence of these multiple worlds.

➢ *Is there any value in Everett's many worlds hypothesis?*

Where do I personally stand on the issue of interpreting quantum mechanics? I used to align myself with the conventional Copenhagen interpretation that, while not denying that there is a real world independent of our observations of it, nevertheless, there is nothing meaningful we can say about it. However, there is a problem with this.

Why do we think that there is a world-in-itself at all? It is because our observations have what we might call an 'otherness' to them. In part we dictate certain features of the observation—for example, whether we are going to measure position or momentum. But having so decided, we do not determine the final outcome—what the actual value of the parameter is going to be. Bohr expressed this idea by pointing out that 'every observation introduces a new uncontrollable element'. Thus we define the world-in-itself to be whatever it is that contributes that 'otherness'—that 'uncontrollable element'—to our observations. Beyond that we can say nothing more about the world-in-itself.

This sounds fair enough until we express exactly the same idea the other way round: we cannot say anything meaningful about what is responsible for the otherness of the observations, but whatever it is, we call it the 'world-in-itself'. In this latter formulation it looks suspiciously as though the expression 'world-in-itself' is little if anything more than a name for our ignorance! But of course we do not solve a problem merely by giving it a name.

It now seems to me that in order for the world-in-itself to be regarded as a meaningful concept, we have to say more about it than simply defining it as that which contributes the otherness of the observations. So we have to ask: is there anything—anything at all—that we can in fact say about it? My response is a guarded yes. I believe that there are a few statements that we can make about its nature. They are of a very general character and not very informative. Nevertheless, they might be adequate for the limited purpose of endowing the concept of a world-in-itself with meaning.

In the first place we note that the otherness of the observations incorporates laws of nature: we discover laws of nature, we do not invent them—they are not part of our input. That law-like behaviour owes its origins to the postulated world-in-itself. That in its turn would seem to imply that the world-in-itself must overall be law-like rather than chaotic. (I include the word 'overall' because from our experience of nature we know that law-like behaviour, in the statistical sense at least, can arise out of a large number of chaotic, random processes. The second law of thermodynamics would be an example of this. As far as we know the law-like quality of the world-in-itself that is responsible for the law-like character of our observations could be of a statistical nature, owing its origins to underlying random processes occurring in the world-in-itself.)

I should make it clear that in describing the world-in-itself as law-like, I do not mean to imply that the same laws that are operative among our observations hold in the world-in-itself. For example, just because observations manifest a conservation law of energy, does not mean that energy is conserved in the world-in-itself, or even that there is such a thing as energy in the world-in-itself. 'Energy' is a word arising out of the domain of observations and might therefore be applicable only to that domain. What I am saying is that there is a certain law-like character of the world-in-itself (whatever form that might take) which corresponds to, or translates itself into, the law of conservation of energy in the domain of observations.

There is also in the otherness of the observations an element of contingency. Today I make an observation known as 'I look in a box and find an electron'. Tomorrow when I look, the electron is no longer there. The manner of making the observation is the same; the box is examined in exactly the same way as before, so my input to the observation is the same. However, the outcome of the observation is different; the otherness of the interaction has changed. This means that there must be something different about the cause of that otherness: the world-in-itself. Let us be clear that the interplay between law-likeness and contingency is a subtle one. In physics we are always examining contingent events, particular particles at particular positions moving with particular velocities at particular times. There is an ever-changing variety of these events, but beneath them all there continuously operate a few simple laws. We arrive at the formulation of those laws through the study of the ephemeral, the transitory, the changing. It is from the contingent that the unchanging laws are distilled. So it must be in the world-in-itself: in addition to law-likeness which remains constant over time, it too must incorporate contingency—the ability to change over time.

Not that this contingency must take the same form as that which appears in the observations, any more than the laws operative in the domain of observations must apply to the world-in-itself. Just because we talk of an electron in a box today and not one there tomorrow does not mean that we are implying that in the world-in-itself there is such a thing as an electron, one that comes and goes. Rather, there is some unknown passing feature of the world-in-itself such that, when it translates its information into the reality domain of observations, it is manifested as an electron. Can we go so far as to say that if we observe two electrons then there must likewise be two contingent features of the world-in-itself, one for each electron? Not necessarily. As we shall have cause to note towards the end of this chapter it might be preferable to think of the observation of two electrons referring to a single feature of the world-in-itself.

Besides law-likeness and contingency another characteristic of observations is their consistency over time. By this I mean that the laws do not change over time. Energy is conserved today and it will be conserved tomorrow and into the indefinite future. We find quarks and electrons today; we expect them to be a permanent feature of our observations. This consistency in the observations must reflect an underlying consistency in the world-in-itself.

The next point to note is that we can alter the world-in-itself; we can have an effect on it. An observation we make now can affect the result of a future observation, and so it has disturbed what is being observed.

Finally we must point out that the world-in-itself is publicly accessible: the world you interact with is the same world that I am interacting with. Not only do you find the same laws as I do but the contingent events tally also. If I observe an electron in a box, then when you look in the box you are likely also to observe an electron. The world you interact with has the same ephemeral features as mine.

There may well be other things that one can say about the world-in-itself, but I think enough has been said to make the point that there appear to be a number of meaningful statements that one can make about it. Admittedly they are of a very general nature only. They are not the kinds of statements that would satisfy those scientists who continue to seek a detailed description of the world outside the context of our observations of it. It is a kind of compromise between on the one hand saying that there is absolutely nothing we can say about the world-in-itself, and on the other saying that there is a world and that it is the job of science to describe it in all its detail.

So there we have it, a quick run-down on the controversy over the interpretation of quantum theory. As we have seen, the issues were hotly debated in the 1930s, and still rumble on today. I suppose it is true to say that most people today incline to one or other version of the Copenhagen interpretation, but the issue was certainly never settled to everyone's satisfaction, and there are still today many prominent physicists deeply unsatisfied with the outcome. And to some extent I sympathise with them. One of the unsatisfactory features of the Copenhagen interpretation is the way it treats as unproblematic the process of making an observation. Implicit in the description of observations is an understanding of the measuring apparatus as behaving in essentially a classical fashion—which seems inconsistent. One thing is certain though: in all the 80 years that the controversy has persisted, no one has come any closer to a detailed description of an objective world out there divorced from the context of making an observation of it. And this is what Bohr would have predicted. We do indeed seem to have come up against a barrier of the knowable.

But there is nothing really new about this. We have already come across the difficulties of trying to say anything meaningful about things-in-themselves, whether it be the nature of time or of space, or the properties of matter such as strangeness and electric charge. What we now find is that not only, like Kant, are we unable to speak of things-in-themselves, but also there is little if anything we can say about the world-in-itself at all!

➢ *Is the task of science to describe the world-in-itself, whether or not it is being observed, or must it confine itself to speaking only of our observations of the world?*

Before leaving quantum theory there is one further feature we need to address: the subject of **entanglement**.

It is a property of certain particles that they possess spin, in much the same way as the Earth spins on its axis. I want you to imagine that we have a particle with zero spin decaying into two other particles, X and Y, which possess spins of equal magnitude. Because of the law of conservation of angular momentum, the resultant spin of the two final particles must be zero like that of the parent particle. In other words, whatever we find as the direction of the spin for particle X we shall find the opposite direction for the spin of particle Y.

Classically when we look at a spinning body—planet Earth, for example—we can find its spin axis pointing in any direction, but with subatomic particles their spin behaves differently. The measured direction of the spin depends on the orientation of the apparatus used for measuring it. In the present case, we use an applied magnetic field to examine the small magnetic moment associated with the spin of the particle. We find that the spin direction of the particle lies either exactly in the same direction as the applied magnetic field, or in the exact opposite direction—it is never at an intermediate angle. It is a case of all or nothing, and never something in between. Thus, the result we get depends in part on how we choose to set up our apparatus— the orientation of the magnet.

Having produced particles X and Y through the decay of the parent particle, we now allow them to separate to a great distance. Eventually they are so far apart that there is effectively no longer any physical force between them. Let us say that you are at the location of particle X, and I am positioned with the distant particle Y. You decide to measure the spin of X.

You set up your detector, choosing at random some particular orientation for the direction of the applied magnetic field. Whatever result you get, you are able to deduce what result I will get if I set up my detector in the same orientation as yours and measure the spin of particle Y; my particle will be spinning in the opposite direction to yours. Let us further suppose that you don't set up your apparatus until after the two particles have separated. My particle will not 'know' in which direction you have chosen to measure the spin of your particle because you did not decide until after the two particles were out of touch with each other. My particle Y must therefore be prepared for all circumstances. It must take with it complete information on what result to supply to me regardless of the orientation of the apparatus. The same kind of reasoning applies to other variables such as, for example, position and momentum. You might decide to measure the precise position of particle X which in turn would allow you to calculate the precise position of my particle Y—which I could then go ahead and verify. Or alternatively you might decide to measure the precise momentum of particle X which would allow you to work out the precise momentum of particle Y, which again I could verify. My particle does not know in advance which of the two—position or momentum—you are going to choose. So, the argument goes, my particle has to be prepared for all eventualities, and be ready to supply me with either the precise position or the precise momentum as required. This is not to say that we can ever measure both—the uncertainty principle forbids this. That is not the point. The point is that it would appear that particle Y does at least have to *possess* all the relevant information on both parameters.

This was the kind of argument advanced by Einstein, Podolsky, and Rosen in a famous paper published in 1935 entitled *Can quantum-mechanical description of physical reality be considered complete?* This so-called **EPR argument** was the culmination of Einstein's famous debates with Bohr. It appeared that the argument proved that each particle possessed locally all the information that one could possibly want. It is just that the uncertainty principle does not allow us to access all that existent information. Therefore quantum theory is not a complete theory, there being information out there in the world-in-itself that we are not accessing. This is not to say that while travelling between the observations the particles necessarily and at all times possess a well-defined spin, or position or momentum. That information might be encoded in hidden variables of some sort. The point is that more information is invested locally in each separate particle than is being extracted. Hence quantum theory is an incomplete theory.

As you might expect, Bohr disagreed. For one thing, Einstein, Podolsky, and Rosen were trying to talk about what was supposed to be happening in between the observations—which Bohr had ruled out as being meaningless. He replied by stating that it was incorrect to think of the two particles as separately constituting two distinct one-particle systems, and that when a measurement is made on one of the particles, the other system is unaffected. He claimed that because the particles had interacted together momentarily in the past they had become entangled. Even though they might appear to be two isolated particles, they are not. We are not dealing with two one-particle physical systems, but a *single* two-particle system. Accordingly an interaction that appears to us to be a measurement on one of the particles is in truth an interaction with the whole two-particle system. Likewise when the second verifying measurement is made, that too is not to be regarded as affecting only one of the particles (the second of the particles this time) but also the entire two-particle system. It was for this reason that I earlier warned you that just because our observation features two electrons, that does not necessarily mean that in the world-in-itself there must be two entities, each corresponding to an observed electron. It could be that the world-in-itself has a single entity that manifests itself in the realm of observations as two separated electrons.

But, you might ask, how can this be? What is it that binds the two particles together as a single system if there is no physical force between them? (Recall how we allowed the particles to separate to some great distance where the gravitational and electric influences between them can be ignored.) There does not seem to be any 'glue'. A further objection is that Bohr's idea entails that somehow information is transferred instantaneously across space from one component of the two-particle system to the other. Surely nothing can travel faster than light. That is true, but only in the sense that no object or form of energy can move faster than light. But nothing physical like that is passing between the particles. It is 'information' that is being transmitted. Einstein was jokingly to dismiss such talk as 'spooky action at a distance'.

And not just Einstein but many people were unhappy with Bohr's seemingly vague, hand-waving way of getting out of the EPR argument. For a time it was thought that there was no way of resolving the issue, but then in 1965 John Bell astonished the physics world by devising a method for deciding between the two competing schools of thought. His idea was as follows: So far we have thought of measuring the spin of one particle and then verifying that the other particle has opposite spin by having the second detector aligned in the same direction as the first. But what if the second detector were oriented at some other angle? What would we expect to get for the spin of the second particle? Instead of the spin always being in the opposite direction, as was the case with the detectors parallel to each other, we would expect the second particle's spin to be sometimes aligned in the opposite direction of the second detector, and sometimes in the same direction. The percentage frequency for getting the same direction, as opposed to the opposite, would depend on the angle between the directions of the two detectors. That much was obvious. But what was not obvious was the fact that Bell was able to show that the frequency also depended on whether one was to regard the two particles as constituting an entangled combined system, as Bohr would have it, or alternatively were to be treated as two separate, independent localised particles, each carrying all the necessary information, in accordance with the EPR type of thinking. In other words, Bell showed that there was an experimental test that could decide between the competing interpretations.

The experiment was carried out in the early 1980s by Alain Aspect and his collaborators using photons passing through polarisers. What did they find? They found that their results favoured Bohr's Copenhagen interpretation. The photons were entangled, forming a two-particle combined system. More recent experiments have confirmed these findings.

So there we have it. Whether we like it or not, we appear to be stuck with this 'spooky action at a distance'—information being transferred instantaneously across a distance. Quantum physics again supplies us with a strong candidate for an unanswerable question—a further encounter with the boundaries of the knowable.

➢ *How are we to understand quantum entanglement, i.e. 'spooky action at a distance'?*

And moreover this gaping hole in our understanding applies to the very simplest multiparticle system one can imagine: two subatomic particles, so far separated from each other that there is no physical force between them, and whose only connection is that, for a fleeting moment at some time in the past, they had been together. Now if that is so for such a simple case, what must entanglement do for the most complicated thing we know in the world—the human brain. Just think of all those particles in close physical contact with each other and possessing a long history of interactions with each other. Surely when it comes to trying to explain the instrument through which we do our thinking, there is an overwhelming call to approach the task with a proper sense of humility.

12

Quantum gravity and string theory

The two pillars upon which is built our modern understanding of the physical world are general relativity and quantum theory. The first is used to describe the large and massive; it is our best description of the cosmos. The latter comes into its own on the small scale, the realm of atomic and subatomic physics. Both have withstood every test in their respective domains.

However, there is a deep worry: the two theories are incompatible with each other. Quantum theory assumes the existence of a background spacetime. It is in that passive arena that all the events we observe going on in the world are played out. General relativity, on the other hand, shows how the contents of the universe shape, mould, and define the spacetime background. There is an intimate interplay between the actors in the drama and the stage upon which they perform. The nature of the background comes out of the theory; it is not fed in a priori.

Not only that, but the whole ethos of each of the two theories is different. Relativity assumes a smoothly changing, continuous, gently curving, spacetime with well-defined properties at each infinitesimal point. All physical variables have well-defined values (as was the case in classical Newtonian physics). Quantum theory, in contrast, is essentially discontinuous: discrete measurements or observations separated by periods of non-observation. According to quantum theory, the very fabric of spacetime is subject to the violent fluctuations caused by the formation of virtual particles on the small scale—fluctuations that become ever more extreme the smaller the scale until eventually it all blows up when one theoretically gets down to a point. As for the physical variables, they cannot all be simultaneously measured or even defined—a consequence of the uncertainty principle.

It is all very well saying that we can just carry on using one theory for the big phenomena and the other for the small. It is aesthetically unpleasing to have two such very different theories based on two incompatible ways of viewing the world. Moreover, there occur in nature situations requiring the two to overlap. Take black holes for instance. The idea of there being black holes arises out of relativity theory. When a supermassive star runs out of fuel, it reaches the end of its active life and can no longer support itself against its own internal gravitational attraction. There is a supernova explosion, with the core of the star sucked down inexorably to a point of no volume at the centre—a singularity. Or so we suspect. This surely takes us into the domain of quantum theory. Again, take the Big Bang. The entire universe emerging from another supposed singularity.

So general relativity has its problems with infinities. Infinite density at the centre of black holes and at the instant of the Big Bang. And quantum theory has trouble with infinities too. These arise as a consequence of the electric and magnetic fields having different values at different points in space—and there are an infinite number of points in space, and thus an infinite number of values for the field to be specified even in a finite volume. So there are an infinite number of variables, and this in turn can lead to the equations going haywire.

What is needed is a combined theory of **quantum gravity**. That is the physicists' Holy Grail. It was a theory that Einstein sought in his latter years—but failed to find.

Currently the best candidate for this, and the one most theoretical physicists have been working on for the past couple of decades is **string theory**. This is a theory that aims not only to unite general relativity and quantum theory into a harmonious whole, but also makes us re-evaluate elementary particle physics. The basic idea is as follows:

In examining in ever greater detail how matter is constructed, we have been probing ever smaller component parts. The expectation has been that eventually we will come to component parts that have no size at all and thus could have no internal structure made up of yet smaller components. That would be a neat way of ending the Russian doll sequence. And as far as we know it could well be that quarks, leptons, and the force-carrying intermediary particles are essentially point-like objects and that there is nothing more to say about them. That has been our assumption so far, but it might not be like that. When we examine the properties of the zoo of particles created in particle accelerators we find systematic similarities between them. It appears as though the higher mass particles are 'excited' versions of lower mass ones. By this we mean that they have the same properties as the lower mass ones, but just possess ever greater internal energies (and hence mass). This gave rise to the idea, developed in the 1970s and culminating in an important paper published by Michael Green and John Schwarz in 1984, that perhaps the truly fundamental particles are not after all point-like particles but strings—tiny infinitesimally thin one-dimensional strings—open strings and closed strings (Figure 23). As such they have a finite length. And yet, despite that finite length, they are not to be thought of as being made up of component parts (unlike familiar household strings). The string itself is the indivisible fundamental entity.

The strings are thought to be under tension and can vibrate at characteristic frequencies, much like a guitar string. Just as we have found that we can produce matter with well-defined allowed values of mass, but nothing between those allowed values, so a guitar string of a particular

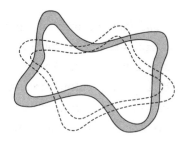

FIG 23 Fundamental particles as vibrating strings.

length and under a particular tension can produce only certain frequencies and nothing in between those harmonics. The succession of evenly spaced peaks and troughs that make up the vibrational mode must exactly fit into the string, and this sets constraints on the allowed values of their wavelength, and hence the note they are sounding. Thus, the different particles we produce in nuclear collisions are merely different vibrational modes of identically similar strings. The properties of the particles, such as their electric charge and colour charge, are to be attributed to different characteristics of the mode of vibration. In particular, the higher the frequency of vibration, the more energy the string has, and hence the more mass. And as there are expected to be an infinite number of different vibrational modes, getting ever more frantic and energetic, so one expects an infinite number of particles—all but a few having masses way beyond what conventional particle accelerators can produce.

This idea applies to all fundamental particles—both the constituents of matter and the force-carrying particles. It is a unifying principle that has given rise to the claim that string theory, when fully developed, will be 'The Theory of Everything'.

How big will these strings be? We cannot tell, but seeing that the whole idea of string theory is to unite general relativity with quantum theory, it would seem natural that there should be some way of deriving string lengths from the fundamental quantities characteristics of those two theories: c, h, and G. Hence our best guess is that the lengths of the fundamental strings will be of the order of the Planck length. This is 10^{-20} times the size of the proton. It is so small that it gave rise to the earlier idea that the fundamental particles were point-like objects. However, small though the length is, it is nevertheless finite, and this finite extent is the bridge between quantum theory and general relativity in that it smears out the quantum fluctuations—it is insensitive to probing what might be happening on a smaller, sub-Planck length scale. In the process, it does away with the problems associated with supposed behaviour at a point. This reson-ates closely with the ethos of contemporary relativity and quantum theory whereby it is held that all that is meaningful is what can be measured. Recall how our inability to measure a well-defined position and momentum for an electron was not due to the clumsiness of the energetic photons we used to observe them with. Rather, it was because a simultan-eously well-defined precise position and momentum for the electron does not exist. In the same way, it is not the case that we are unable to investigate the catastrophic fluctuations as one approaches point-like distances because we are using clumsy string-like objects to examine them with. Rather, the whole idea of catastrophic singularities does not arise at all because it was erroneously based on the notion of point-like objects, to say nothing of spatial extents still being meaningful even below the Planck length. This involved the supposed singularity at the instant of the Big Bang and also those that were supposed to exist at the centre of black holes. According to string theory there is no such thing as a singularity.

The theory is attractive. After all, physicists do not know how to handle point-like singularities, and any theory that neatly gets rid of them is to be welcomed. It is for this reason, as was said earlier, that most theoretical physicists have for more than two decades been working on string theory.

Initial signs were good. It was found that the vibrational modes of the string necessarily included one that corresponded to the characteristics expected for the proposed graviton. Thus, the supposed mediator of the gravitational field seemed to come out of the theory in a natural way. Other features of the Standard Model also emerged in a natural way.

Unfortunately, string theory also gave rise to vibrational patterns that corresponded to particles having imaginary mass, i.e. particles the mass of which when squared was a negative quantity. In point of fact, soon after special relativity was first formulated, there had been speculation as to whether, in addition to ordinary matter that was confined to move with speeds between zero and the speed of light, c, there might be a family of other particles that were constrained always to move with speeds between c and infinite speed. Such hypothetical particles, called **tachyons**, would indeed have imaginary masses. There is no evidence for their existence, so their prediction on the basis of string theory was an embarrassment. However, it was found that this could be overcome by incorporating supersymmetry into string theory. You will recall that we earlier mentioned the possibility of the known particles having supersymmetric partners. They were in fact first suggested in the context of string theory. Nowadays all string theories are of the supersymmetric type. Indeed, vibrational patterns corresponding to the two types of partner seem to come out of the theory in a natural way.

The characteristics of the vibrational modes depend on the number of different spatial dimensions in which the vibrations can take place. If the world had been such that there was only one spatial dimension, then the only kind of vibration possible would be of the back-and-forth type exhibited by a slinky coil along its length. The existence of a second dimension allows us to add a component of up-and-down motion. In three dimensions we have even greater freedom. The addition of each new dimension opens up fresh possibilities. It is found that in order to reproduce all the features of fundamental particles (their masses and charges), six more spatial dimensions have to be added to our familiar three.

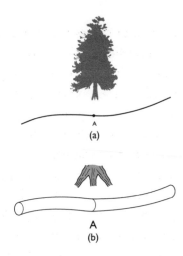

FIG 24 A hosepipe as seen (a)
from a distance, and (b) close up.

Which is all very well, but where are these supposed
additional dimensions?! We do not see them. To counter
this objection, string theorists suggest that the additional
dimensions are curled up small. As an analogy, consider a
situation where you are observing someone's garden from a
distance. You see a line of some sort snaking across the
lawn (Figure 24a). You might think that in order to specify a
point on the line one would need just one number—the
distance of the point from the end of the line. On getting
closer to the garden you discover that the line is actually a
hosepipe (Figure 24b). Now you realise that at any distance
from the end of the pipe there is not a uniquely defined
point but a circular cross-section through the pipe. To
define an actual point on the hose you would need not
only the distance to the end of the pipe but also a measure

215

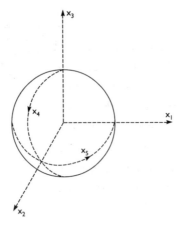

FIG 25 Two of the curled-up dimensions that are postulated to exist at each point in the more familiar three-dimensional space.

of its distance round the circumference of the cross-section from some chosen reference point. Thus string theory says that in spacetime you need not only the usual three spatial and one time measurement to define the location of an event, but also six other numbers to locate the event along the curled-up dimensions that exist at each point in space (Figure 25). It is the number of dimensions, their sizes and shapes, that determine the masses and charges of the particles.

Why are there three extended spatial dimensions and six curled up ones? No-one knows and perhaps we shall never know. It is conjectured that perhaps it is due to some random historical occurrence—how things happened to have developed in the very earliest stages of the Big Bang as a result of initial conditions, the details of which have now been lost.

> *How many dimensions are there? And why are some curled up, and others extended?*

One of the advantages of string theory was held to be the fact that, unlike the Standard Model with its 19 adjustable parameters, string theory required only one parameter—the tension in the string. Equating the characteristics of one particular vibrational mode to those expected of the graviton yields a value of 10^{39} tons force for the string tension. The strings are *very* stiff! This tension causes the strings to be very short—calculations indicating a length of the order of the Planck length.

One adjustable parameter? That sounds good. Unfortunately, there's a catch. We saw how supersymmetry is indispensably incorporated into string theory, but it turns out that there are five ways of doing this. And these alternatives differ in other ways, for example, whether they include strings that are open and have two loose ends, or are closed loops. There is not a unique theory; we are left with a choice, and no way of deciding between them.

In fact, matters get worse when we discover that there is a vast choice of different geometries for the curled-up dimensions. Thus, the number of unknown quantities has gone up rather than down. There is a whole 'landscape of theories'. Different geometries lead to different vibrational modes, which in turn lead to different predictions for the properties of the particles. Only the prediction of a particle with the expected characteristics of a graviton is common to all. There seems no way of deciding which is the one to go for—if any of them.

But then in 1995 there came a landmark conjecture by the American physicist Ed Witten. He suggested that there might be a truly fundamental theory—a meta-theory—provisionally called **M-theory** (M standing for 'meta' or perhaps for 'mother', 'mystery', or 'membrane'), and this unique theory spawns the five main variants that have currently caught the attention. The cost of such a theory is that it appears to introduce yet another dimension. This comes about because in M-theory, the strings are not actually one-dimensional strings but two-dimensional membranes that extend into the hitherto unrecognised extra spatial dimension. This yields a total of 11 dimensions: 10 spatial dimensions and one of time. So in summary, we have gone from point particles to strings, and from strings to membranes. But this is as yet pure speculation. In any case, it does not address the problem as to why one particular string theory it spawns is the one to be physically realised in the actual world.

One possible answer to this is again to call upon the multiverse hypothesis. Accordingly one postulates that *all* the string theories spawned by M-theory go on to be realised somewhere in an infinite number of universes. Again pure unverifiable speculation. No-one knows what form the M-theory takes—nor even whether there is an M-theory.

➤ *Is there an M-theory, and if so, what is it?*

In fact, it has to be said that string theory in general lacks good sound predictions—the kind where one can go away, do an experiment, and confirm or refute the prediction. After all, that is the way science usually progresses. String theory has so far made no predictions—or it only makes predictions that cannot be verified (extra dimensions invisible because they are rolled up, or particles not seen because they are too heavy).

It can be argued that this assessment is not quite fair. As we have seen, the characteristics of the graviton emerge so naturally out of string theory that one might be justified in claiming that string theory 'predicts' gravity. It is just a matter of historical accident that we happen to have become experimentally acquainted with gravity before we came across its prediction by string theory. Not only that, but one can also say unequivocally that string theory requires particles to have supersymmetric partners. None of the known particles qualify in this regard to be some other particle's partner. It is suggested that this might be because all the superpartners are so heavy that we have so far not been able to create any of them. Perhaps the new high energy LHC at CERN will find them. If it does, that may well be thought of as a genuine prediction arising out of supersymmetric string theory. But

219

even so, that in itself will not constitute proof of string theory's validity. It is perfectly possible to have supersymmetry with point-like particles.

Another inconvenient feature of string theory is that not only are the extra dimensions too small to be seen, but so are the strings themselves—of the order of the Planck length. In order to see them one would need projectile beams with a wavelength smaller than the string—otherwise one would not be able to see the details. It would be a physical impossibility to build an accelerator capable of probing to such distances; it would have to be the size of a galaxy.

In view of all this, there is a growing unease that, in the absence of further genuine predictions that can be subsequently tested, the early promise of string theory might be illusory. This has even given rise to the suggestion among certain fervent string theorists that perhaps our understanding of what science is should be modified so that experimental verification of predictions should not always be a requirement. Some theories are so attractive that they should be accepted anyway. This seems a dangerous path to go down. The history of science is littered with aesthetically pleasing theories that have had to be subsequently abandoned in the face of experimental evidence to the contrary—the steady state theory and the closed universe hypothesis, to name but two. This whole question of point-like particles or strings may well turn out to be unsolvable. We shall have to wait and see.

> *Is there any way of proving the validity of some form of string theory?*

But in any case, is string theory, as currently for-mulated, a basis for quantum gravity? The trouble is that string theory, like quantum theory itself, assumes the existence of a spacetime background. The strings move about in a passive spacetime, just as the quantum measurements take place against a passive spacetime background. They are both background-dependent. By contrast, general relativity is background-independent; the spacetime background emerges out of the theory. In general relativity, par-ticles do not move about in a passive fixed space. The characteristics of the space are affected by all forms of energy present. The particles have energy, so at the same time as they move through the space they are contributing to the characteristics of that space.

Thus it is expected that a theory of quantum gravity is likely to start off from a quantised basis, and the features of spacetime will emerge out of it. This contrasts with our normal procedure where we start off with the assumption of a space and then put things in it to move about. For spacetime to be quantised presumably means that there might be a smallest division of spacetime. As we have said before, perhaps the idea of geometry, and of size, loses all meaning below something like the Planck length. Putting together all these quantised 'bits', the illusion of a smooth spacetime emerges.

So, string theory as presently formulated points the way towards a theory of quantum gravity, but cannot itself be that theory until it is somehow rendered background-independent.

Will we ever be able to formulate a fully satisfactory theory of quantum gravity? ≺

As for it being 'The Theory of Everything', the most that can be said is that it might form the basis of a theory of all *physical* phenomena. If one is a reductionist, then one holds that the physical is all that there is, and hence the theory does indeed cover everything. But if one believes that even a complete theory of the physical world would still not account for everything—consciousness, free will, aesthetics, morals, the spiritual—then such a claim is without foundation.

Does complete understanding require more than solely physical explanations? ≺

13

Concluding remarks

So ends our tour of the major problems facing fundamental science today. The questions found listed in the margin do not constitute *the* definitive list; there is no such thing. Other writers might have come up with a somewhat different list. I have doubtless overlooked some questions that deserved to be included. But at least what you have before you gives you some sense of where we have got to, and what the remaining challenges are.

For readers without much in the way of a formal training in science I can well imagine that the going was a bit tough at times. If so, my apologies. The subject matter of this book is such that it does indeed at times stretch our human intelligence just about as far as it will go. All of us, professional scientists included, can find it demanding when confronted with ideas that at first seem counter-intuitive. But at least I hope you now have some feel for the fascination of working at the frontiers of scientific knowledge.

Which of those listed questions are likely to remain with us for all time? I have indicated what seem to me to be the intractable ones—the questions that take us right up to the boundaries of the knowable. But as I said at the beginning, these are merely expressions of my own personal opinion—for what they are worth. In some instances I might be proved wrong. I only hope I am still around to find out. Indeed it is difficult to see how one could ever be *absolutely* sure that one had come up against the boundaries of the knowable in all directions. One thing does, however, seem certain to me: fundamental science must have its limitations. The scientific age in which we live can be but a transitory phase in human development. At some stage it must come to an end. We are deeply privileged to find ourselves living in this special time—a time of discovery, innovation, and the exploration of human capabilities.

INDEX